Peterson First Guide to Reptiles and Amphibians

HOUGHTON MIFFLIN COMPANY

Boston New York

Illustrations and text based on *Field Guide to Reptiles and Amphibians Eastern and Central North America,* 3rd ed. (1991) by Roger Conant and Joseph T. Collins and *Field Guide to Western Reptiles and Amphibians* (1985) by Robert C. Stebbins

Library of Congress Cataloging-in-Publication Data

Conant, Roger, 1909–

Peterson first guide to reptiles and amphibians/
Roger Conant, Robert C. Stebbins, Joseph T. Collins.
p. cm.
ISBN 978-0-395-97195-6
1. Reptiles—North America—Identification. 2. Amphibians—North America—Identification. 3. Reptiles—North America—Pictorial works. I. Stebbins, Robert C. (Robert Cyrill), 1915– .
II. Collins, Joseph T. III. Title.

QL651.C66 1992
597.6'097--dc20 91-33016
CIP

Printed in China

SCP 16 15 14 13

4500386549

Editor's Note

In 1934, my *Field Guide to the Birds* first saw the light of day. This book was designed so that live birds could be readily identified at a distance, by their patterns, shapes, and field marks, without resorting to the technical points specialists use to name species in the hand or in the specimen tray. The book introduced the "Peterson System," as it is now called, a visual system based on patternistic drawings with arrows to pinpoint the key field marks. The system is now used throughout the Peterson Field Guide series, which has grown to over 40 volumes on a wide range of subjects, from ferns to fishes, rocks to stars, animal tracks to edible plants.

Even though Peterson Field Guides are intended for the novice as well as the expert, there are still many beginners who would like something simpler to start with—a smaller guide that would give them confidence. It is for this audience—those who perhaps recognize a crow or a robin, buttercup or daisy, but little else—that the Peterson First Guides have been created. They offer a selection of the animals and plants you are most likely to see during your first forays afield. By narrowing the choices—and using the Peterson System—they make identification even simpler. First Guides make it easy to get started in the field, and easy to graduate to the full-fledged Peterson Field Guides. This one gives the beginner a start on the reptiles and amphibians.

Roger Tory Peterson

Introduction

Time was when the only good snake was a dead one. Fortunately, as we have come to understand that every species has a place in the global environment, that attitude is almost a thing of the past. We now know that the fear of reptiles and amphibians is not instinctive, but is learned by children, usually from people who are simply uninformed. The fact is that many of these animals make excellent neighbors because they eat rodents or insects.

What is a reptile? What is an amphibian?
Both reptiles and amphibians are cold-blooded, meaning they depend on the sun or other heat source to stay warm. Beyond that, there are several differences between the two groups. Reptiles—the snakes, turtles, lizards, and crocodilians—have scales or plates, and their toes have claws. (The clawless Leatherback sea turtle is an exception.) Young reptiles are miniature versions of their parents.

Amphibians—the salamanders, toads, and frogs—have moist skins, and most have no claws. Their young have a larval stage, usually passed in the water (such as the tadpole of a frog) before they change into their adult form. In fact, the word *amphibian* is based on Greek words meaning "living a double life."

How to use this book
This First Guide uses the system of identification invented by Roger Tory Peterson for his Field Guides. Arrows on the illustrations point to key markings or other characteristics unique to the animal, and important parts of the description are printed in *italics.*

Remember that no two individuals are exactly alike. Those that have wide ranges, especially, often look very different in various parts of the country. In some cases, we have included illustrations for more than one form.

The measurements given in the descriptions are for the entire animal, including its tail. Many professional herpetologists (scientists who study reptiles and amphibians) do not include the tail. In the case of lizards, for example, the tail is often missing or shortened, so it is more reliable to measure the animal from snout to vent (or anus). But we feel that including the tail in the measurement often gives a better idea of the real size of the animal.

Making the catch

Except for the turtle basking on a sunny log, most reptiles and amphibians are hiders. They are not likely to be seen at all unless you go hunting for them. Even those that prowl are apt to be mere streaks in the grass. So you will usually have to catch the animal before you can identify it. When hunting for reptiles and amphibians, look under things. Rocks, logs, and even trash on the ground are all good hiding places.

Your success at finding animals will vary with the weather. Wet weather is the time for amphibians. After heavy rains, a sunbaked area that has been powder-dry for months may swarm with hopping frogs and toads. Reptiles are most likely to be found when the weather is warm.

Many kinds of reptiles and amphibians reflect light from their eyes, like a cat. Go out at night with a flashlight and watch for their eyeshine, holding the light near your head. Dewdrops and the eyes of spiders glint silver and green, those of moths and toads yellow or red. You can also see eyeshines when driving slowly at night on little-traveled roads. The eyeshine method works best with toads, frogs, and turtles. Eyeshines of snakes and salamanders are too faint to be seen well, and lizards are chiefly active in the daytime.

Your own two hands are the best tools for catching reptiles and amphibians. Most kinds can just be grabbed, but you had better be quick about it, for they are adept at slipping away. *Never try to catch a venomous snake.* Learn to recognize the dangerous species instantly, and stay well away from them.

A harmless snake can be caught by stepping on it *gently* and pinning down its head with a stick. It's a good idea to wear gloves when capturing large ones. Pick up the serpent by grasping it firmly but gently behind the jaws.

A slapping-down technique can be used for lizards and frogs that freeze rather than dashing away. Slap your flattened hand (gently!) over the animal and pin it down while you grasp a leg with your other hand. You will find it easier to approach your quarry if you don't look directly at it. Move in at an angle, watching it from the corner of your eye. And remember never to grab a lizard by the tail, for it may break right off in your hand.

Care in captivity

After you catch a snake or frog or turtle, what do you do with it? Usually you will want to hold it briefly, check its identity, admire its color, perhaps take its picture, and then let it go. But you may wish to keep it for a while. Some reptiles and amphibians are easy to keep in captivity, and they are fun to watch. But they should be liberated *where you found them* within two or three weeks.

Most reptiles and amphibians can be kept in an aquarium. Make sure it is big enough for your guest and that it has a tight-fitting lid. You can make a lid with a wire screen tacked to a wooden frame.

To make a home for many amphibians, lizards, and small turtles, turn the aquarium into a terrarium, with a layer of pebbles covered with clean soil and some small plants. Water the plants two or three times a week. When they are sprinkled, the lizards will lap up the drops. (Many will not drink from a dish.) Frogs and toads will need a shallow dish of water.

Snakes usually don't do well in terrariums, for many of them burrow. Many toads and turtles also like to dig. These animals are best kept on bare pebbles, but shelters may be added. Hiding places are a necessity. You might use large pieces of bark or a plastic container with a door cut through it.

Keep your captives warm. When they are cold, they may stop eating. Too much heat, though, can kill them. Temperatures of 75° to 85° are best for most reptiles, but place the heat source only at one end of the aquarium. Amphibians thrive in cooler temperatures, usually below 70°.

Most reptiles and amphibians eat either meat or insects. Turtles may be fed canned dog food that is mostly meat, and should also have some leafy vegetables such as spinach. Box turtles should have greens, soft fruits, and berries, as well as some meat.

Toads, frogs, and most lizards and salamanders need live insects. Some will starve to death even if surrounded by platoons of dead bugs. Drop in a live one and it will be seized at once.

Many snakes have specialized feeding habits. Make sure you know what to feed any snake you try to keep.

Before you decide to keep any reptile or amphibian, check with the authorities in your state or region—usually the Fish and Game or Wildlife Department. They can give you a list of the species that are protected by law in your state.

Conservation

Many kinds of reptiles and amphibians are rapidly disappearing from areas where they were once abundant. Habitats are destroyed at a great rate so the land can be used for human activities; many animals are senselessly slaughtered by uninformed people who kill anything that moves; and some species are too much in demand by collectors. You can help.

Be kind to habitats. Replace stones, logs, and other hiding places that you overturn in your searches.

Never collect rare or endangered animals, and refuse to buy them from anyone else. Concentrate on studying reptiles and amphibians in the field. Practice with a camera and make your collection a photographic one. Such a collection is permanent, needs no cleaning or feeding, and is not destructive to wild populations.

SALAMANDERS

Moisture is an absolute necessity for these amphibians. Even those that live on land can survive only in damp environments.

Giant Salamanders

These big, bizarre salamanders look more like bad dreams than live animals. Despite their fearsome appearance, the three shown here are completely harmless.

MUDPUPPY **8–19 in.**

Most salamanders have gills only in their larval stage, but the Mudpuppy keeps its *external gills* all its life. The gills are like miniature red ostrich plumes waving gracefully in the water. The Mudpuppy has a dark stripe through its eye and usually has some dark spots. In the South, this salamander is often called a "waterdog." Contrary to belief, Mudpuppies do not bark. They are found in lakes and rivers from Canada and the Great Lakes to Louisiana.

GREATER SIREN **20–38 in.**

This giant salamander can be over three feet long. It looks like an eel with *little front legs* and *external gills*, which are crowded together near its head. The legs are so small they may be hidden by the gills. Greater Sirens live in shallow-water habitats such as ponds and ditches, in Florida and along the Atlantic Coast up to Maryland. Because they live in the water and are active only at night, few people ever see them.

HELLBENDER **11–29 in.**

This huge, grotesque, and quite slimy salamander has a flattened head and a wrinkled fold of fleshy skin along each side of its body. Adults do not have external gills. Usually lives in running water, from New York southwest to Arkansas. Hellbenders may be found by slowly overturning large rocks in clear streams. They are widely but mistakenly thought to be poisonous.

MUDPUPPY

GREATER SIREN

HELLBENDER

Mole Salamanders

Like moles, these amphibians stay underground most of the time, often in the burrows of other animals. But many kinds gather in large numbers in pools and ponds after early spring rains to mate and lay eggs.

SPOTTED SALAMANDER **$4^{1}/_{2}$–$9^{1}/_{2}$ in.**

This woodland salamander has up to 50 *round yellow* or *orange spots* in *two uneven rows* on its back, from eye to tail tip. During the warm rains of early spring, many Spotted Salamanders migrate into ponds to breed. Look for them under stones or boards in damp places or during wet weather. They live mainly in the Northeast, but also south to Georgia and Texas.

TIGER SALAMANDER **7–13 in.**

The *light spots* or *"tiger stripes"* of this salamander vary depending on its locality, but usually they are whitish or yellow on a black background. The markings are *irregular* in shape and don't form a row. Tigers used as fish bait are often accidentally introduced into new habitats by anglers. They are widespread through many parts of the U.S.

MARBLED SALAMANDER **$3^{1}/_{2}$–5 in.**

This chunky salamander has *white* or *gray bands* that may enclose *large dark spots.* The female lays her eggs in a low spot that will be filled by the next good autumn rain, and she may stay with the eggs until they are covered with water. Lives in the East, from New England to Texas.

LONG-TOED SALAMANDER **$3^{1}/_{2}$–6 in.**

This salamander lives in many different habitats, from dry sagebrush plains to mountain meadows, in the Pacific Northwest and western Canada. It is dark with a tan, yellow, or olive-green stripe on its back. The stripe may be *broken up* into a series of spots. A form with yellow-orange spots and bars that lives in northern California is *endangered* and should not be caught.

SPOTTED SALAMANDER

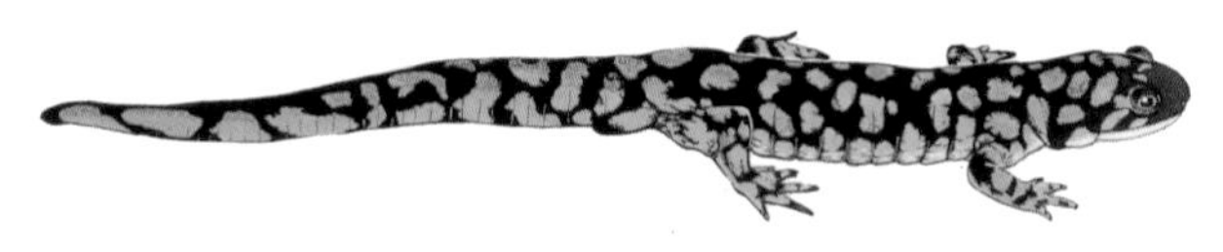

TIGER SALAMANDER

MARBLED SALAMANDER

LONG-TOED SALAMANDER

Newts

Newts are not as slippery as most other salamanders. Their skins are rougher and not slimy, and they don't slide through your fingers when you try to handle them. Most newts live in the water, but the young of many kinds go through a land stage.

EASTERN NEWT **2–5½ in.**

The larvae of this salamander transform into *red efts*, small land animals that remain ashore for one to three years. Then they return to the water and change into aquatic adults. The most brilliantly colored efts live in moist, forested mountains from Nova Scotia to east Texas. They are quite bold and may be found walking about the forest floor. Adults live in clean ponds, marshes, and lakes. They may have red spots. Adult males grow high tail fins and horny growths on their hind legs when they are ready to mate.

CALIFORNIA NEWT **6–8 in.**

In their land stage, this newt and its relatives, the Rough-skinned and Red-bellied newts, are easily distinguished from other western salamanders by their *rough skins*. The breeding males, found in ponds, streams, and reservoirs, has smooth skin and a flattened tail. Newts are brown or black above, yellow or orange below. The California Newt has *pale lower eyelids*. It lives in the mountains of California.

EASTERN NEWT

CALIFORNIA NEWT

Lungless Salamanders

These salamanders "breathe" through their moist, slippery skins. They are the largest family of salamanders. Most live on land, under rocks, logs, and other hiding places, but in the East many species live in and near streams. The salamanders on pp. 14–23 all belong to this family.

RED SALAMANDER **4–7 in.**

The Red Salamander has many *small black spots* on its upper surfaces, and *yellow eyes.* Young adults are generally brilliantly colored, and older ones are darker. One form has a black chin. Look for it under moss and stones in springs and streams, even trickles, with water that is clear and cool. It lives in streams that run through meadows, fields, and woods. The Red Salamander ranges over a large part of the eastern U.S., from New York to Mississippi and Louisiana.

MUD SALAMANDER **3–8 in.**

A rusty or red-colored salamander, with or without round *black spots.* It has *brown eyes* but is otherwise very much like the Red Salamander. To tell the two species apart, make note of the habitat: the Mud Salamander usually lives up to its name, preferring mucky places. Found mainly in the southeastern states, from New Jersey to the Gulf Coast.

SPRING SALAMANDER **4½–9 in.**

Most often a cloudy orange or salmon color with a *black mottled pattern* or spots. Prefers cool springs and wet forest areas in the East, from Maine south through the Appalachian Mountains and into Alabama. Turn over stones in mountain brooks to find these brightly colored amphibians.

RED SALAMANDER

MUD SALAMANDER

SPRING SALAMANDER

The group of salamanders shown here is often abundant in humid forests. When it is damp or rainy, they prowl at night and can be observed with a flashlight.

REDBACK SALAMANDER **2–5 in.**

The Redback is the most common salamander in most parts of its range, which extends from southern Quebec and Minnesota to North Carolina. There are two distinct color forms. The *redback* has a reddish stripe down its back; the *leadback* is all dark gray to black. Both kinds have a mottled *salt-and-pepper* belly.

WESTERN REDBACK SALAMANDER **2–5½ in.**

The *straight-edged stripe* down the back of this salamander extends to the tip of its tail. The stripe may be tan, reddish, orange, or yellow. Some Western Redbacks are plain orange or pale yellow. Lives near the Pacific Coast in British Columbia, Washington, and Oregon. Look for it under damp rocks, boards, and logs in humid forests.

NORTHERN SLIMY SALAMANDER **4½–8 in.**

Slimy Salamanders are mostly all black above, with scattered *silvery white flecks* and spots. They are well protected from predators by sticky skin secretions that cling to your hands like glue and almost have to wear off. A variety of Slimy Salamanders are found in most parts of the eastern half of the United States, usually in woodland ravines and hillsides.

JORDAN'S SALAMANDER **3½–7 in.**

The humid forests of the southern Appalachian Mountains are the home of this extremely variable salamander. The form shown is black with *red cheeks;* another has *red legs;* and still others are plain black or have white spots on their sides. Its habitats include leaf litter on the forest floor, rotting logs, and mossy stone piles.

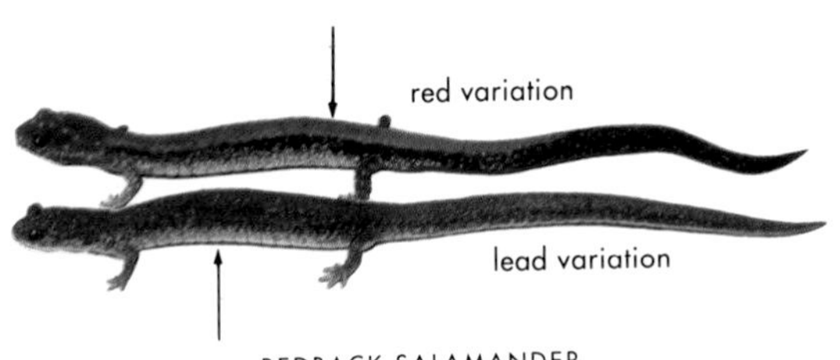

REDBACK SALAMANDER

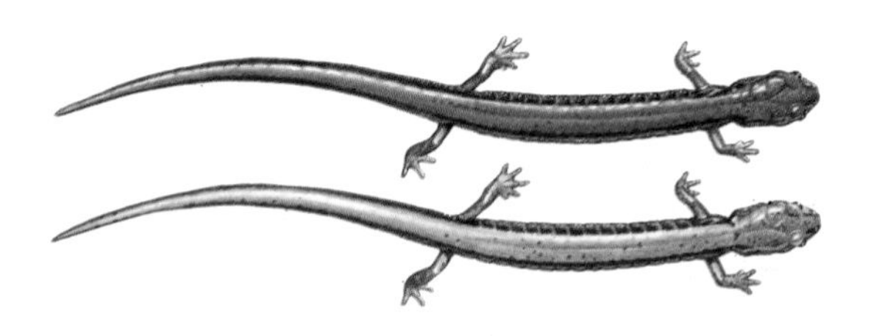

two variations

WESTERN REDBACK SALAMANDER

NORTHERN SLIMY SALAMANDER

JORDAN'S SALAMANDER

The salamanders on these two pages live in brooks, springs, and other small bodies of water where there are usually no fish, which might eat salamanders. Most of them are yellow in color, at least on their undersides.

CAVE SALAMANDER **4–7 in.**

This salamander's favorite habitat is in the twilight zone of caves, near entrances where the light is dim. It is an excellent climber, and moves about on the rocks and ledges, sometimes hanging on only by its very long tail. The Cave Salamander is orange or reddish, with black spots. It lives in areas where there are abundant limestone caves, from Virginia to Oklahoma.

LONGTAIL SALAMANDER **4–7½ in.**

A yellowish salamander with black herringbone or "dumbbell" shaped markings on its tail. The tail makes up more than half its total length. The Longtail may wander about on land in warm, damp weather, but it also is a good swimmer. Found from New York to Alabama and eastern Missouri, often in caves.

NORTHERN TWO-LINED SALAMANDER **2½–4½ in.**

This is the common *yellow* salamander of the Northeast. The *dark lines* down the sides may break up into dots or dashes on the tail. It is a brookside salamander, hiding under all sorts of objects at the water's edge and running or swimming away vigorously when alarmed. In warm, wet weather it may wander far out into nearby woodlands.

CAVE SALAMANDER

LONGTAIL SALAMANDER

NORTHERN TWO-LINED SALAMANDER

DUSKY SALAMANDER **$2^1/_2$–$5^1/_2$ in.**

The Dusky's color varies depending on where you find it within its range in the eastern United States. It is generally gray or brown, with markings not much darker. It may have several pairs of *golden* or *yellowish spots* down its back. The tail has a *keel* (a slender ridge) along its upper side, and the body is short and stout, with hind legs larger than the forelegs. Dusky Salamanders are talented jumpers, and can leap several times their own length to escape being caught. They are seldom found far from running water and are most common along edges of small streams where they can hide under stones or debris.

CALIFORNIA SLENDER SALAMANDER **3–$5^1/_2$ in.**

A slim salamander with very *short legs* and a very *long tail.* Often has a stripe down its back that may be red, brown, tan, or yellow. Usually lives in woodlands, grasslands, even yards and vacant lots. Females lay their eggs in late fall or winter, often in nests shared with other females, and the young hatch in winter and spring. Lives mainly near the coast of California.

ENSATINA **3–6 in.**

Here is another often blotchy salamander; this one lives on the West Coast from British Columbia to Mexico. Ensatinas come in many colors, from plain brown above to brightly marked forms that are dark with orange or yellow blotches or orange bands. When touched, Ensatinas stand ready to defend themselves, stiff-legged and with tail arched. In cold or dry weather they hide underground, often in animal burrows.

DUSKY SALAMANDER

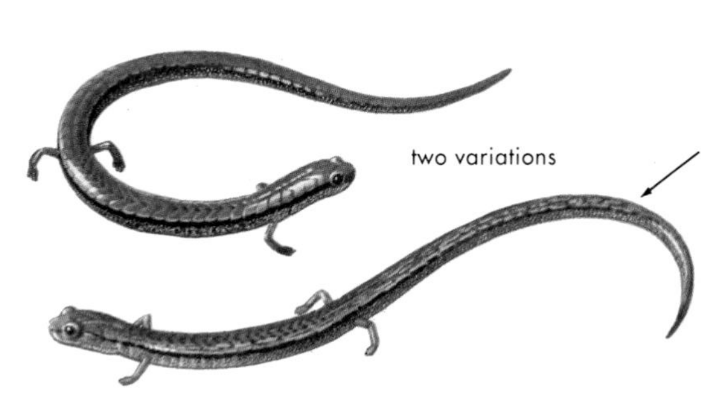

CALIFORNIA SLENDER SALAMANDER

ENSATINA

ARBOREAL SALAMANDER **4–7 in.**

Search for this climbing salamander at night by looking under things or examining tree trunks and rock walls with a flashlight. A brown salamander, usually *spotted with yellow.* Males have a chunky, triangular head with powerful jaw muscles. Usually squeaks when caught, and large adults have been known to bite hard. Found in the coastal mountains and valleys of California. In hot and dry weather, Arboreal Salamanders gather in tree hollows and damp caves and mine shafts.

GREEN SALAMANDER **3–5½ in.**

Here is an eastern climbing salamander. It may be seen in trees, but it is more often a cliff dweller. It is usually found in narrow crevices of damp Appalachian rock faces that are protected from direct sun and rain. Its flattened head and body help it get around in tight places. It has square-tipped toes. This is the only really *green* salamander in North America, and it has darker green markings that look like lichens.

MOUNT LYELL SALAMANDER **3–4½ in.**

This *web-toed* salamander is an expert rock climber. On slopes, the Mount Lyell Salamander uses its short tail like a walking stick: the tail is rhythmically curled forward and its tip placed against the ground. It is found chiefly among the granite rocks of the Sierra Nevada of California. Look for it under rocks near cliffs, cave openings, melting snowbanks, and in the spray of waterfalls. It can be hard to spot, because its coloration closely matches granite. Its flat head and body help it to maneuver in cracks and crevices.

ARBOREAL SALAMANDER

GREEN SALAMANDER

MOUNT LYELL SALAMANDER

FROGS AND TOADS

There are no hard and fast rules for distinguishing between toads and frogs. In general, though, the typical toad has a warty skin and short legs for hopping, and the typical frog has a smooth skin and long legs for leaping. Both are found on land and in water, but toads can easily live far from water.

Many male toads and frogs advertise their presence with their voices, especially at breeding time. Each kind has its own mating call, and some also make other noises, such as the "scream" of a captured frog, rain calls, and various warning chirps and squeaks.

In dry areas, these amphibians may estivate, a process like hibernation that helps the animals survive times of drought.

True Frogs

These are the typical frogs. They generally have long legs, narrow waists, and smooth skin.

BULLFROG **3½–8 in.**

This is our *largest* frog. It is aquatic and likes bigger bodies of water than most other frogs. Listen for its call, a loud, bass-noted *jug-o'-rum* in its breeding season. Bullfrogs are easy to find at night with a flashlight, by their eyeshine. When first seized, a Bullfrog may "play possum," hanging limp and motionless, but watch out for sudden recovery! Bullfrogs are native to the eastern U.S. and have been introduced into many western areas. In some of its new homes, in fact, it is considered a pest, because this big amphibian has a voracious appetite and will eat almost anything that moves. Once established, it can wipe out local populations of other species of frogs and small turtles and snakes. It is usually green above, but in the Southeast the Bullfrog may have a dark pattern, and individuals from Florida are almost black. Look for the conspicuous *eardrum*, in males much larger than the eye.

BULLFROG

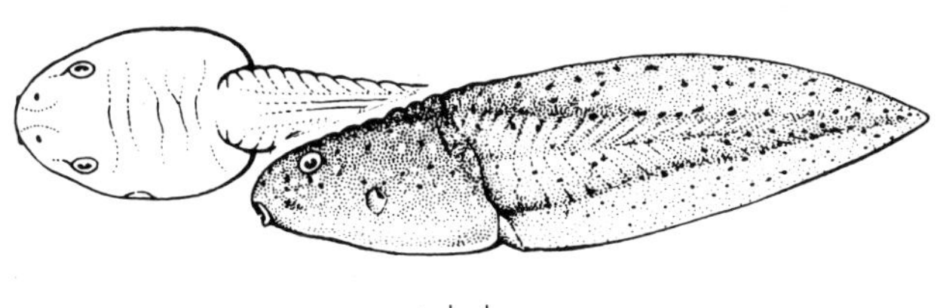

tadpole

CARPENTER FROG **$1^1/_2$–$2^1/_2$ in.**

This frog gets its name from its voice, which sounds like two carpenters hitting nails a fraction of a second apart: *pu-tunk, pu-tunk.* It often lives in or near sphagnum bogs, where it may be difficult to find because its brown color blends so well with the acid waters of the bogs. It may be seen at the water's surface with only its head exposed. The Carpenter Frog has two light stripes down each side. It lives in the eastern Coastal Plain from southern New Jersey, including the Okefenokee Swamp of Georgia and Florida.

GREEN FROG **2–4 in.**

Despite its name, the color of this frog depends on where you find it. In most of its range in the eastern U.S. it is green or brown, but in the South it is usually bronze with a mostly white underside. In males, the eardrum is twice the size of the eye. Though often abundant, it is a secretive frog, hiding in logs and swamps or rock crevices. It lives wherever there is fresh, unpolluted water—springs, creeks, ponds, and ditches. Its call is a twanging baritone note, *clung* or *c'tung,* like a loose banjo string.

WOOD FROG **$1^1/_2$–3 in.**

The frog with the *robber's mask.* You can always see the dark patch extending back from the eye, though the body color of this frog varies considerably. Even the same individual may at different times be pink, brown, or almost black. It lives in the northeastern U.S. and in much of Canada, in moist wooded areas. The Wood Frog's voice is a hoarse clacking sound that suggests the quack of a duck.

CARPENTER FROG

two variations

GREEN FROG

WOOD FROG

RED-LEGGED FROG **2–5 in.**

A western frog with *red* coloring on its lower abdomen and the *underside of its hind legs*. It is chiefly a pond frog, preferring sites where cattails or other plants provide good cover. It is usually found near water but may wander after a hard rain into deep woods and meadows. Its voice is a stuttering series of notes on one pitch, ending in a growl: *uh-uh-uh-uh-uh-rowr.* Calls are sometimes given from under water. The Red-legged Frog lives near the Pacific Coast from Canada to Baja California. At the turn of the 20th century it was the source of many of the frogs' legs sold—and eaten—in California.

FOOTHILL YELLOW-LEGGED FROG **1½–3 in.**

This frog is *yellow* on the underside of its legs. Its body color may be gray, brown, reddish, or olive. Its colors usually harmonize with the color of the rocks and soil around it. Most often found along streams, especially where there are riffles and rocks and sunny banks. Its voice, seldom heard, is a guttural, grating sound. Lives near the coast in Oregon and in the mountains of California. This frog has become rare in much of its range.

RED-LEGGED FROG

FOOTHILL YELLOW-LEGGED FROG

PICKEREL FROG **2–3½ in.**

A frog with *square spots* arranged in two parallel rows down its back. There is bright yellow or orange on the hidden surfaces of its hind legs. The Pickerel Frog has a wide range that includes most of the eastern U.S. It likes cool, clear water in the North, and in the South it occupies the warm, often tea-colored waters of Coastal Plain swamps. Often found in the twilight zone near the mouths of caves. Few snakes will eat Pickerel Frogs, probably because the secretions of their skin glands make them distasteful. Their voice is a steady, low-pitched snore. Males often call while submerged under water.

NORTHERN LEOPARD FROG **2–4 in.**

A brown or green frog with two or three rows of *round spots* with light borders on its back, as well as many smaller spots on its sides. Its range extends from most of Canada south as far as New Mexico. It is found in a variety of habitats—grassland, brush, and forest, ranging high into the mountains. In summer, the Northern Leopard Frog wanders far from water into meadows. Its mating call is a long, deep, rattling snore, somewhat like a motorboat. Leopard Frogs have long been used as laboratory animals.

SOUTHERN LEOPARD FROG **2–5 in.**

Similar to the Northern Leopard Frog, but it usually has a distinct *light spot in the center of its eardrum* and has only a few dark spots on the sides of its body. It is found in all sorts of freshwater habitats, and even enters slightly salty marshes along coasts. Its mating call is a short, chuckling trill. The range of the Southern Leopard Frog extends from Long Island south to the Florida Keys and west to Texas.

PICKEREL FROG

NORTHERN LEOPARD FROG

SOUTHERN LEOPARD FROG

MINK FROG **2–3 in.**

The skin produces an odor like the scent of a mink (some say rotten onions) when the frog is rubbed or handled roughly. It has various dark spots on its back and dark blotches on its hind legs. Its voice is a deep, burred *cut-cut-cut-cut-cut.* The Mink Frog lives in the far North, along the borders of ponds and lakes or the cold waters of streams. Look for it where water lilies are plentiful; it likes to venture far out from shore by hopping from pad to pad.

CRAWFISH FROG **2–4½ in.**

A stubby-looking frog whose dark spots are *encircled by light borders.* Its color varies, depending on the temperature, the frog's activity, etc. It is often found in crawfish holes that have lost their "chimneys" and contain water. Other habitats include the burrows of small mammals, holes in roadside banks, and storm sewers. Its voice is a loud, chuckling snore. Lives in the central U.S. from Iowa and Indiana to eastern Texas.

Tailed Frogs

This family of frogs, named for its tail-like reproductive organ, has only one member.

TAILED FROG **1–2 in.**

A flat-bodied frog that looks rather toadlike, with its rough skin and olive or brown coloring, which matches the surrounding rocks. Usually has a pale yellow triangle on its snout and a *dark eyestripe.* Its *outermost* toe is the *broadest.* It prefers clear, cold, rocky streams in the forests of the Pacific Northwest. The tadpoles cling to rocks with their suckerlike mouths. Only the male has a "tail."

MINK FROG

CRAWFISH FROG

TAILED FROG

Chorus Frogs

Many people hear Chorus Frogs, but few ever see them. During their breeding time, they sing night and day near shallow bodies of water, hiding in clumps of grass or other vegetation so well that they are extremely hard to find even when they advertise their presence by calling loudly.

SPRING PEEPER **1 in.**

The Spring Peeper has a *dark cross* on its back, usually in the form of an X. These small singers form their choral groups where trees or shrubs are standing in the water, or at least nearby. The voice is a high, piping whistle. A large chorus of Spring Peepers heard from a distance sounds like sleigh bells. Widespread in the eastern and central U.S.

ORNATE CHORUS FROG **1 in.**

Looks like the creation of an imaginative artist. The colors of this little frog are quite variable — an individual may change from almost black to silvery white or to the bright colors shown in the illustration, including the *bold black spots* on its sides. The voice of the Ornate Chorus Frog is a series of bird-like peeps, or like the ring of a chisel struck by a hammer. It lives in the southeastern Coastal Plain from North Carolina to Louisiana, and calls during late fall, winter, and early spring.

STRECKER'S CHORUS FROG **1–2 in.**

This largest and chubbiest of the Chorus Frogs lives in the central U.S., mainly in Texas and Oklahoma. The *stout hand and forearm* are quite toadlike. Its general color is variable, but it has a masklike stripe from snout to shoulder and a *dark spot* below the eye. The voice of the Strecker's is clear and bell-like, but in a large chorus the effect is of a rapidly turning pulley wheel badly in need of greasing.

SPRING PEEPER

two variations

ORNATE CHORUS FROG

two variations

STRECKER'S CHORUS FROG

LITTLE GRASS FROG **Less than 1 in.**

The tiniest frog in North America. Most people mistake it for a baby frog of some other species. Though its general color varies, the Little Grass Frog almost always has a dark line through its eye and extending onto the side of the body. This elfin Chorus Frog's climbing is restricted to low-growing plants. It favors the grassy environs of ponds of the Southeast, from Virginia to Florida. Its tinkling, insectlike call is so high and shrill that some people have difficulty hearing it.

WESTERN CHORUS FROG **1–1½ in.**

A small, slim frog, usually with *three dark stripes* down its back. Though it needs shallow bodies of water at breeding time, otherwise this frog survives in a wide variety of habitats. It has adapted well to the presence of humans and even lives in suburbs and near cities, except in areas polluted by pesticides. This frog, in all its variations, is found over a very wide area, from the southern Atlantic and Gulf coastal states to the Rocky Mountains, and north in western Canada almost to the Arctic Circle. You can imitate its *prreep, prreep* by running a finger over the smaller teeth of a pocket comb.

PACIFIC CHORUS FROG **1–2 in.**

Also called the Pacific Treefrog. A small frog with a *dark stripe* through its eye. It is usually green but may be brown, reddish, or gray. An individual can change from dark to light in a few minutes, but its basic color does not vary. This is the most commonly heard frog on the Pacific Coast. Its call is a loud, two-parted *kreck-ek,* rising on the second syllable. This Chorus Frog lives in many different habitats, from sea level into the mountains—grassland, forests, even desert oases.

LITTLE GRASS FROG

WESTERN CHORUS FROG

PACIFIC CHORUS FROG

Treefrogs

These frogs are well adapted to life in the trees. Their toes are equipped with adhesive pads, and their long legs and toes help them cling to twigs. Many Treefrogs can change color, and the same frog may be gray at one time, brown at another, either patterned or plain colored. Baby Treefrogs may be bright green.

PINE BARRENS TREEFROG **1–2 in.**

This beautiful little frog has purplish stripes bordered by white and set against bright green. It is a resident of the swamps, bogs, and brown, acid waters of the New Jersey pine barrens, the bogs of the Carolinas, and extreme western Florida. Its call is a nasal *quonk-quonk-quonk.*

BARKING TREEFROG **2½ in.**

One of the larger, stouter Treefrogs—and the spottiest. The spots are usually visible even when the frog changes color, but they may disappear when the frog turns dark brown or bright green. The Barking Treefrog is both a climber and a burrower, and lives in many different habitats in the Southeast. Its barking call of nine or ten raucous syllables is delivered from high in the treetops. The breeding call, given near water, is a single explosive *doonk.*

GRAY TREEFROG **1–2 in.**

A moderately large Treefrog found throughout the eastern half of the U.S. It has a *light spot* below its eye. Normally gray or gray-green, but individuals may vary from brown or green to almost white, depending on changes in their activity and environment. Bright orange markings are concealed in the folds of the hind legs. The Gray Treefrog's voice is a musical trill that some people think sounds like the call of the Red-bellied Woodpecker.

PINE BARRENS TREEFROG

BARKING TREEFROG

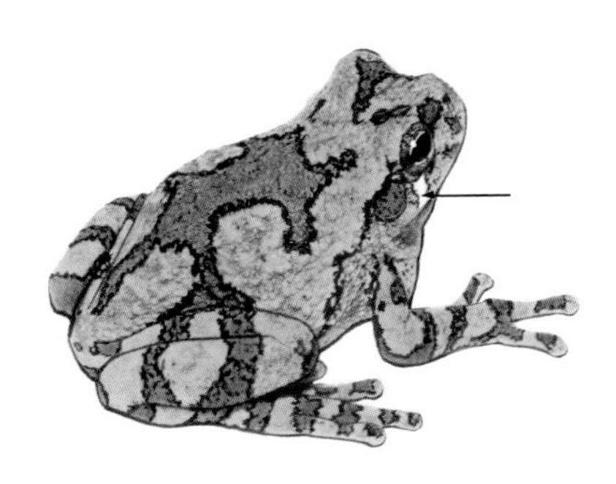

GRAY TREEFROG

PINE WOODS TREEFROG **1–1½ in.**

This little frog is an arboreal acrobat that climbs high in the trees, though it is often found at lower levels. It is usually a deep reddish brown color, but at times may be gray or greenish gray. There is a row of small orange, yellow, or whitish spots hidden behind the thigh, but you cannot see these when the animal is at rest. It lives in the southeastern Coastal Plain. The Pine Woods Treefrog has a "dot-and-dash" voice, like Morse code done with a snore. A large chorus sounds like a lot of riveting machines all operating at once.

SQUIRREL TREEFROG **1–1½ in.**

Like a "chameleon" in its myriad variations of color and pattern. The same frog may be brown or green, plain or spotted. Usually there is a spot or dark bar between the eyes. It lives in the South, where it is a ubiquitous animal that appears suddenly in and around houses, even seeming to drop from the sky as it falls out of a tree while acrobatically pursuing insects. Lives almost anywhere it can find moisture, food, and a hiding place. Its voice is like a duck's, but more nasal.

CANYON TREEFROG **1–2 in.**

This Treefrog occasionally climbs trees, but it prefers the ground. It lives in the boulder-strewn canyon creeks of the arid Southwest and Mexico. Brown, cream, or olive-gray, it is well camouflaged as it huddles in crevices on the sides of rocks within easy jumping distance of the water. Its voice is a series of short, nasal notes—*ah-ah-ah-ah*—that sounds as if the frog were calling from inside a tin can.

PINE WOODS TREEFROG

two variations

SQUIRREL TREEFROG

two variations

CANYON TREEFROG

SOUTHERN CRICKET FROG **1 in.**

A southern and lowland frog. It has a *dark triangle* between its eyes, a pointed head, and long hind legs. The Southern Cricket Frog lives in bogs and ponds and river-bottom swamps in the southeastern Coastal Plain, mainly from Virginia to Louisiana. Its voice is like a metal clicker—*gick, gick, gick.*

NORTHERN CRICKET FROG **1 in.**

A northern and often upland frog, with a blunter head and shorter hind legs than the Southern Cricket Frog. Also has a *dark triangle* between its eyes. It ranges from New York to Texas. Its voice is like two pebbles being clicked together, slowly at first and then picking up speed.

Tropical Frogs

Most of the members of this large family live in the American tropics, but a few range north into Texas and the Southwest.

GREENHOUSE FROG **1 in.**

A tiny immigrant, probably from Cuba. It has two different patterns: *striped*, with longitudinal light stripes, or *mottled*, with irregular dark and light markings (both are shown). It is generally brown or reddish. Greenhouse Frogs live on land, hiding by day under boards, leaves, or even trash where there is some moisture. Often found in greenhouses, gardens, dumps, or stream valleys. Its voice is a series of melodious, birdlike chirps. It was introduced in Florida and is now widespread there.

BARKING FROG **2–3½ in.**

A frog that looks like a toad but has a smooth skin (no warts). It is tan or greenish colored. It lives mainly in Mexico, but its range extends north into limestone caves and ledges in Texas and New Mexico. The Barking Frog rarely ventures out into the open. When captured, it puffs itself up alarmingly. Heard from a distance, its explosive call sounds like the bark of a dog.

SOUTHERN
CRICKET FROG

NORTHERN
CRICKET FROG

GREENHOUSE FROG

BARKING FROG

Toads

The homely "hoptoads" are chunky and short-legged, with rather dry, warty skins, and they hop, instead of leaping like frogs. You will not get warts from touching toads, but their skin secretions will irritate tender areas like your mouth and eyes. Wash your hands after handling toads, and don't touch your face until you do.

OAK TOAD **1 in.**

Our tiniest toad, with a conspicuous *light stripe* and four or five pairs of dark spots on its back. Some of its warts are red or orange. The Oak Toad is abundant in the pine woods of the South, from Virginia to the Florida Keys and Louisiana. Its voice is like the peeping of newly hatched chicks.

GREEN TOAD **1–2 in.**

The bright green color, small black spots, and flat head and body make the Green Toad easy to identify. It lives in rather dry habitats and is rarely seen abroad except during and after heavy rains. Its voice is a shrill trill, somewhat like a policeman's whistle but not nearly as loud. Lives in the Southwest, from Kansas south into Mexico.

RED-SPOTTED TOAD **1½–3 in.**

A small toad of the West with a flat head and body and a pointed snout. It is gray, olive, or reddish brown, with *reddish or orange warts.* Its habitats include desert streams and oases, open grassland, and rocky canyons. It climbs among rocks with ease. Voice is a long, musical trill.

GREAT PLAINS TOAD **2–4½ in.**

Our only toad with *large dark blotches,* each with a bold light border. Ground color is gray, brown, greenish, or yellowish. Its shrill, metallic trill can go on for 30 seconds or more. This toad lives in the "wide open spaces" and the dry Southwest. It is a talented burrower.

OAK TOAD

GREEN TOAD

RED-SPOTTED TOAD

GREAT PLAINS TOAD

WESTERN TOAD **$2^1/_2$–5 in.**

You can identify this toad by the *white* or *cream-colored stripe* down its back. It is a dusky or greenish color above, with warts set in dark blotches. The Western Toad uses a wide variety of habitats, from desert springs to mountain meadows. It ranges from Alaska south to Baja California and east to the Rockies. Like many toads, the Western Toad buries itself in loose soil or hides in the burrows of other animals, and it tends to walk rather than hop. Voice is like the peeping of a chick.

YOSEMITE TOAD **2–3 in.**

This toad lives in the high meadows and forest borders of the Sierra Nevada of California, emerging soon after the snow melts. It is similar to its relative the Western Toad but has smoother skin and very close-set eyes. Males and females look quite different; females have many blotches on a pale background, and males are nearly plain yellow-green or olive.

SONORAN DESERT TOAD **4–$7^1/_2$ in.**

All toads have glands behind their eardrums that contain a sticky white poison. This one's poison is powerful enough to paralyze or even kill the unlucky dog or other predator that tries to seize it. The Sonoran Desert Toad is our largest western toad. It is dark brown, olive, or gray above, with smooth skin. Several *large warts on the hind legs* and a *large round white one* behind the mouth are conspicuous against the smooth skin. Also known as the Colorado River Toad. The largest part of its range is in Arizona and Mexico.

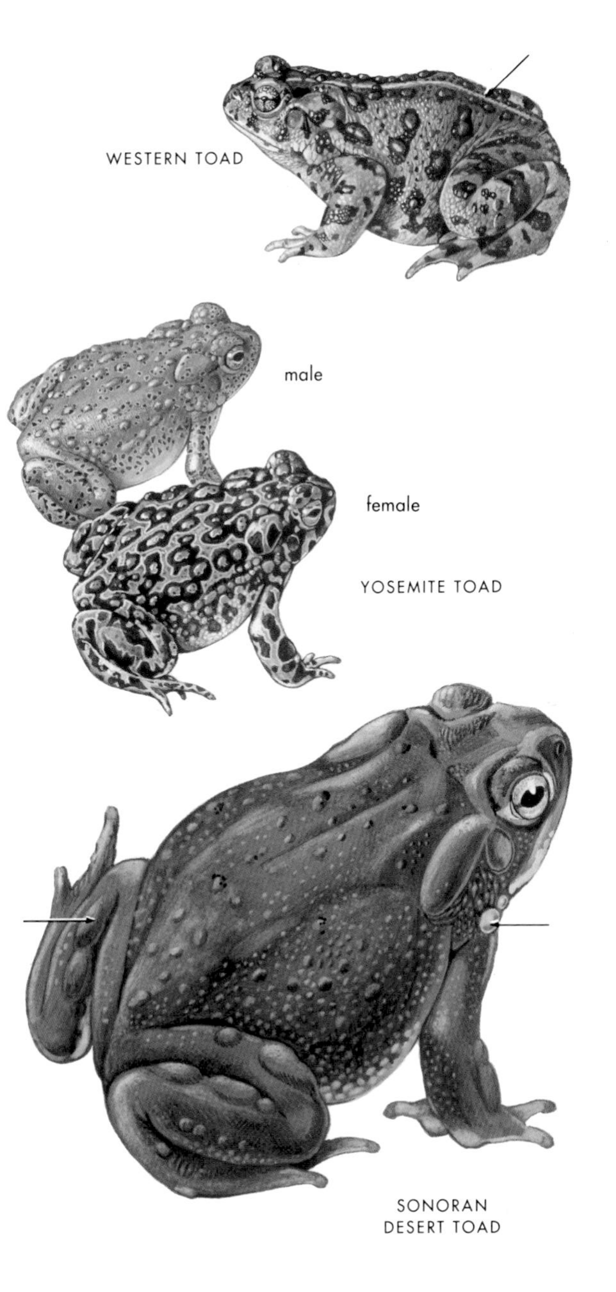
WESTERN TOAD
male
female
YOSEMITE TOAD
SONORAN
DESERT TOAD

WOODHOUSE'S TOAD **2–5 in.**

A large toad with a *light stripe* down its back and dark spots or blotches that contain three or more warts. It is generally brown or gray. Its call, a nasal *w-a-a-a-h,* sounds like the bleat of a sheep or a baby's cry. Woodhouse's Toad lives in many environments, including backyards, over most of the U.S. It usually appears at night, often near lights where insects can be had for the gulping.

AMERICAN TOAD **2–4 in.**

The widespread and familiar "hoptoad." Many are plain brown, but others, especially females, are gaily patterned. Usually has a spotted chest and only one or two warts in each large spot. The American Toad's habitats are legion, from suburban backyards to mountain wildernesses. It is welcomed by gardeners for its boundless appetite for cutworms and other insects. The mating call of the male, a musical trill, is one of the pleasantest sounds of early spring. Ranges from Labrador to Manitoba and south almost to the Gulf of Mexico.

Narrowmouth Toads

These small, plump amphibians have short limbs, pointed heads, and a fold of skin across the back of the head.

GREAT PLAINS NARROWMOUTH TOAD **1–1½ in.**

A smooth-skinned toad with an oddly shaped body and almost no pattern. It may be found at night feeding at anthills. A nearby chorus sounds like a swarm of bees. Occurs from Mexico north to Nebraska.

EASTERN NARROWMOUTH TOAD **1–1½ in.**

Has the distinctive shape of a narrowmouth toad—short legs, broad waist, and pointed head. Its color varies through gray, brown, or reddish, and the same frog may change color. Belly strongly mottled. Its voice sounds a little like an electric buzzer. Lives throughout the southeastern U.S.

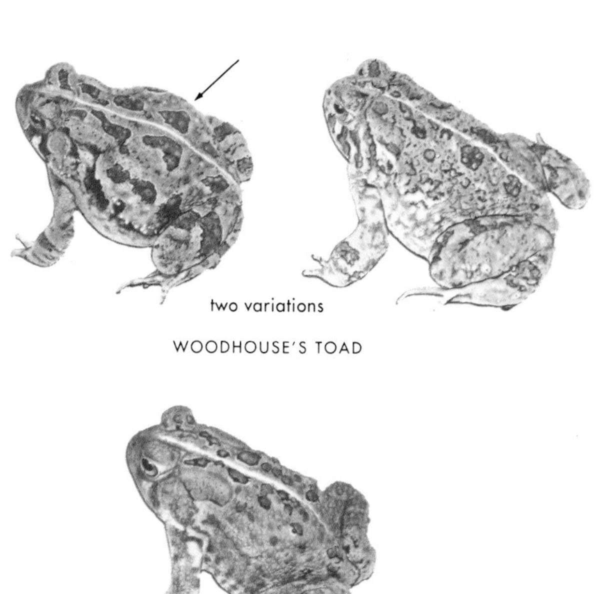

two variations

WOODHOUSE'S TOAD

AMERICAN TOAD

GREAT PLAINS NARROWMOUTH TOAD

EASTERN NARROWMOUTH TOAD

Spadefoots

A small, sharp-edged "spade" on each hind foot enables a spadefoot to burrow straight down and disappear into the earth. It spends most of its time underground, but after heavy rains, large numbers of spadefoots may suddenly appear, heading for breeding ponds. A parched region can reverberate with their mating cries after rains begin.

COUCH'S SPADEFOOT **2–3½ in.**
A spadefoot with black or dark green marbling on a *yellow* ground color. Like other spadefoots, its eye is catlike, having a vertical pupil in bright light. Occurs in the southwestern U.S. and Mexico, most often on short-grass plains and other dry areas. Its groaning bleat suggests a goat or sheep unhappy at being tied.

PLAINS SPADEFOOT **1½–2½ in.**
A spadefoot with a *boss*, or bump, between its eyes. It is generally brown or gray with a greenish tinge, with darker markings. At home on the Great Plains and other regions of little rainfall. It prefers open grasslands and usually avoids wooded areas.

EASTERN SPADEFOOT **2–3 in.**
The only spadefoot east of the Mississippi River. *Two yellowish lines* begin at the eyes and run down the back, forming an hourglass pattern. Lives in the East, in areas with loose, sandy soil.

WESTERN SPADEFOOT **1½–2½ in.**
Dusky green or gray above, often with four irregular light stripes down the back. Skin warts are often tipped with orange. The eye is pale gold, with a vertical pupil. Lives in California, usually in lowland areas such as river floodplains, but also ranges into the foothills and mountains. Its voice is a hoarse snore; a distant chorus is like the sound of a handsaw cutting wood.

COUCH'S SPADEFOOT

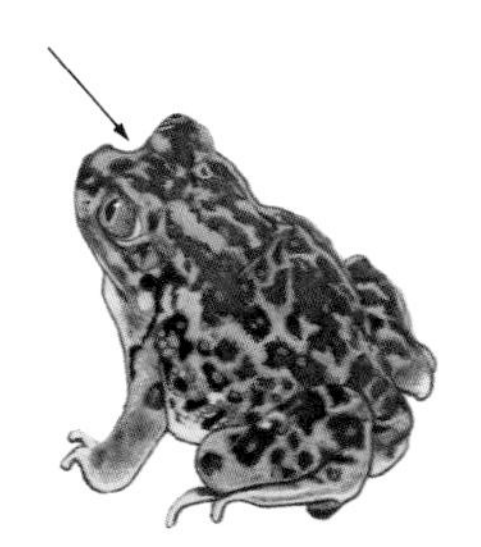

PLAINS SPADEFOOT

EASTERN SPADEFOOT

WESTERN SPADEFOOT

CROCODILIANS

These huge, lizardlike reptiles live in many of the warmer parts of the world. The two species shown here are native to the United States. A third, the Spectacled Caiman, has been introduced in Florida and is now common in some southern parts of the state.

AMERICAN CROCODILE **7–15 ft.**

The *long, tapering snout* is the hallmark of the Crocodile. Its general color is gray. A large tooth in the lower jaw is clearly visible (at close range) when the mouth is closed. Crocodiles are now rare in the U.S. They may be found in Everglades National Park, Biscayne Bay, and the Florida Keys. If you see a crocodilian in salt or brackish water, it is probably a Crocodile. This fearsome reptile will not normally attack humans, but wounded ones are very dangerous. Crocodiles can fight savagely. Males have a voice, a low rumble or growl.

AMERICAN ALLIGATOR **6–19 ft.**

The *broad, round snout* distinguishes this big reptile from the American Crocodile. Overall color is black. Alligators live in fresh water, in the great river swamps, bayous, and marshes of the Gulf and Coastal Plains states, from North Carolina to Texas and throughout Florida. You may spot them basking out of water. Also, watch for eyes, heads, or snouts above the surface at gator holes. Beware of females guarding their nests, which are mounds of plant matter up to seven feet across and three feet high. When the eggs hatch, the female carries the hatchlings gently in her mouth to the water. The voice of the male Alligator is a throaty, bellowing roar with great carrying power. The female roars, too, and grunts like a pig when calling to her young. Alligators of all sizes hiss. In the past, the Alligator was widely overhunted, but under government protection it is now recovering in many areas.

AMERICAN
CROCODILE
young

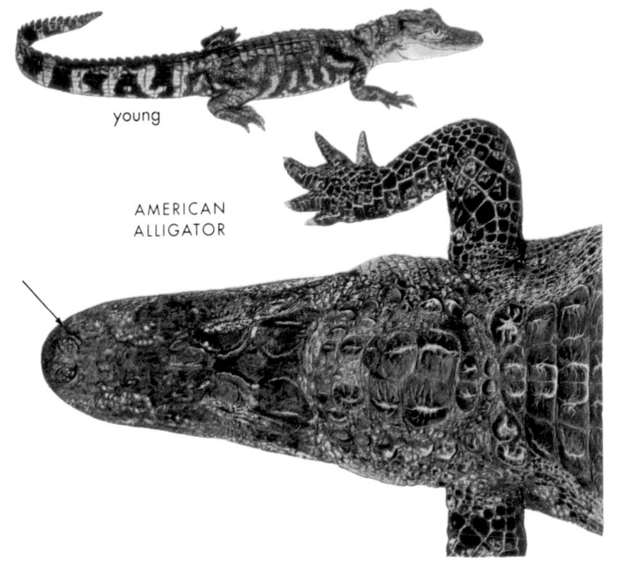
young
AMERICAN
ALLIGATOR

TURTLES

Turtles have survived almost unchanged for 200 million years, but now many of the world's 240 species are threatened with extinction. They can be found from the desert to the ocean, but even those that live in the sea lay their eggs on land. Most are carnivorous, but some will also eat plants. Turtles mature slowly and live long, some to over a hundred years.

Snapping Turtles

Large freshwater turtles with a long tail, large head, powerful hooked jaws, and a small lower shell (plastron). Unable to retreat completely into their shells, they will bite more readily and ferociously than most turtles if they feel threatened.

SNAPPING TURTLE **8-19 in.**
The only snapper in most of North America. Found from the Atlantic to the Rockies, in marshes, ponds, rivers, and streams. Often buries itself in the mud, sometimes in shallow water with only its eyes showing. Rarely seen on land except when laying eggs, usually in early summer. Adults range from almost black to light brown. The young are usually darker, with light spots on the shell. This snapper can be told from the Alligator Snapper by its *saw-toothed tail.*

ALLIGATOR SNAPPING TURTLE **15–31 in.**
The largest freshwater turtle in North America, weighing up to 300 pounds. Found from north Florida to east Texas and up the Mississippi Valley to Iowa and Kansas. Often lies at the bottom of a lake or river with its mouth open, attracting fish with a curious pink growth on the floor of its mouth that wriggles like a worm. Its head is even larger than that of the common snapper, its *shell is rougher,* and its tail is not saw-toothed on top. Young Alligator Snappers have a very long tail.

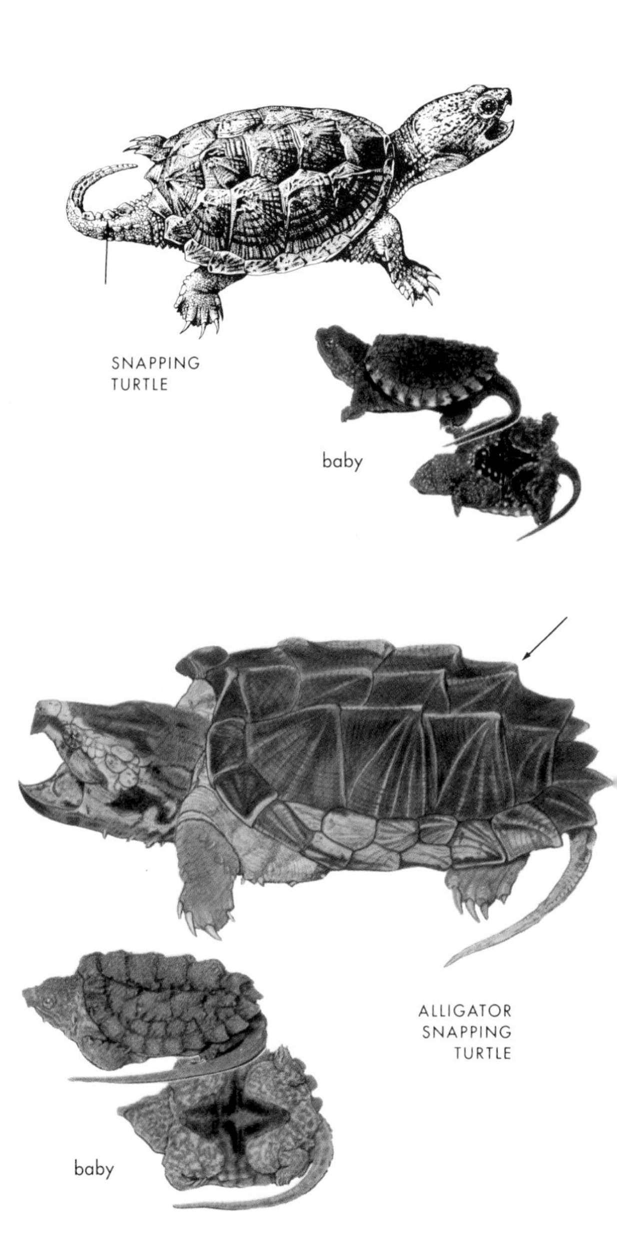
SNAPPING
TURTLE
baby
ALLIGATOR
SNAPPING
TURTLE
baby

Musk and Mud Turtles

These are the "stinkpots" and "stinking jims" that often take a fisherman's hook. Their names arise from the musky odor they give off when caught. Drab in color, they are often mistaken for snappers but have smaller tails and smoother shells. Most kinds seldom leave the water except when nesting.

COMMON MUSK TURTLE **2–5 in.**

If a turtle ever falls on your head or into your canoe, it will probably be a Musk Turtle. With a very small lower shell, it is agile for a turtle and will sometimes climb slender trees to bask in the sun. This turtle is not often seen except in shallow, clear lakes and rivers, where it may be observed patrolling the bottom in search of food, its shell looking like a rounded, algae-covered stone. Has *two light lines* on the side of its head. Found from New England to Florida and west to Texas and Wisconsin.

EASTERN MUD TURTLE **3–4$^{1}/_{2}$ in.**

The lower shell of a Mud Turtle is larger than a Musk Turtle's and is *hinged at both ends*. Prefers shallow water. Ranges from Long Island to the Gulf.

Softshells

These animated pancakes belie the traditional slowness of the turtle. Powerful swimmers, they can also run on land with startling speed. The shell is soft and leathery, the neck long, snout pointed, beak sharp. Handle with care: Softshells can inflict a painful bite.

FLORIDA SOFTSHELL **11–25 in. (female)**

Prefers lakes, while other Softshells usually live in rivers. The largest North American Softshell, with the smallest range: South Carolina to southern Florida.

SPINY SOFTSHELL **7–18 in. (female)**

Has *spines* on the forward edge of the upper shell. Several different kinds are found in rivers from the Great Lakes to the Gulf Coast and in other areas.

COMMON MUSK-TURTLE

EASTERN MUD TURTLE

underside

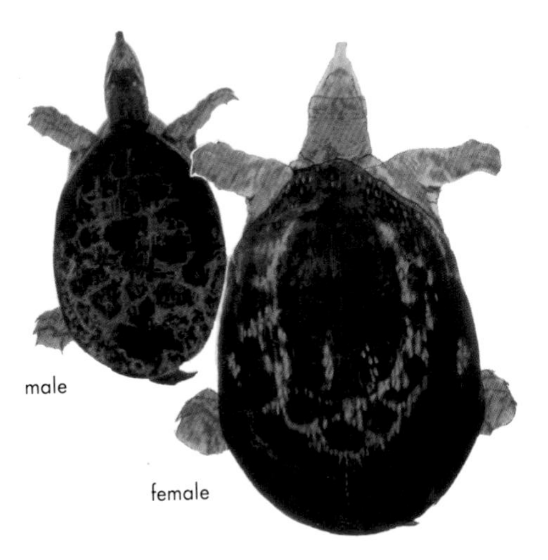

FLORIDA SOFTSHELL

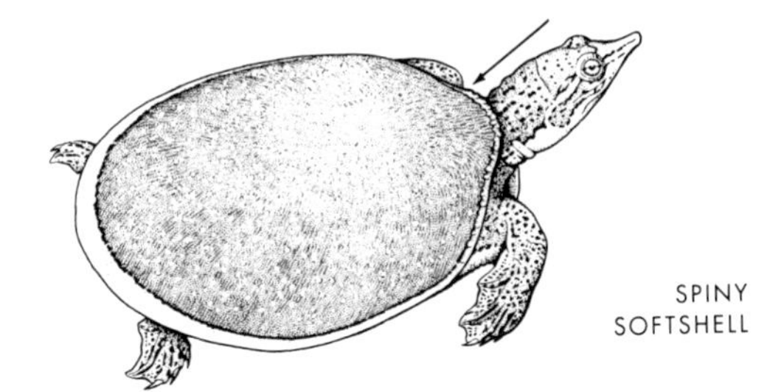

SPINY SOFTSHELL

Pond Turtles

Found chiefly in the Northeast, with one species in the West. Other members of the family live in Europe, Asia, and North Africa.

WOOD TURTLE **$5^1/_2$–9 in.**

The "sculptured turtle." Each large scute of its upper shell rises like a pyramid. Once called "redlegs" for the *orange on its neck and legs*. Like the Box Turtles, usually found on land, where it often wanders long distances. The Wood Turtle is also at home in water, however. Ranges from Nova Scotia to Minnesota and south to Virginia.

WESTERN POND TURTLE **$3^1/_2$–$7^1/_2$ in.**

A water turtle, found not only in ponds but in weedy lakes and marshes. The low, dark upper shell is usually covered with a *network of brown or black markings*. Found along the West Coast from Washington to Baja California.

SPOTTED TURTLE **$3^1/_2$-5 in.**

Black upper shell with *yellow or orange spots*. Prefers shallow water. Seldom in a hurry. If disturbed while basking in the sun it usually enters the water at a leisurely pace and buries itself in the mud. Ranges from Maine west to Illinois and south to Florida.

Terrapins

Turtles of coastal salt marshes, once prized as a food and hunted almost to extinction. Females are larger than males.

DIAMONDBACK TERRAPIN **4-9 in.**

Upper shell varies in color from light gray to black, has *concentric rings* or ridges. Lower shell can be orange to greenish gray, sometimes with dark markings. Legs and neck are spotted. Several different varieties of Diamondbacks can be found along the Atlantic and Gulf coasts from Cape Cod to Texas, in almost any sheltered and unpolluted body of salt or brackish water.

WOOD TURTLE

WESTERN POND TURTLE

SPOTTED TURTLE

DIAMONDBACK TERRAPIN

Cooter, Slider, Redbelly, and Painted Turtle

Along with the Map Turtles, these are the most common and conspicuous of the basking turtles. They often bask in the sun for hours on rocks or logs in or beside a pond or stream. Sometimes one will climb on top of another until they are stacked two or three deep.

PAINTED TURTLE **4–10 in.**

The smooth dark upper shell has *red markings* around the edge and lacks the ridgelike keel down the middle that some other turtles have. The lower shell is usually yellow, often with dark blotches. The head and throat are *streaked or spotted with yellow.* Four different types of Painted Turtles are found, collectively, from coast to coast in the northern states and as far south as Louisiana and Alabama. The variety shown here lives in the north-central U.S. and west to Washington and British Columbia.

RIVER COOTER **9–17 in.**

The dark shell has *yellow markings*, often in the shape of concentric circles. Like most basking turtles, River Cooters slip into the water at the least sign of danger. Several kinds of Cooters and River Cooters are found in different parts of the South. They differ widely in appearance; the eastern variety is shown here.

SLIDER **5–11 in.**

Usually has a *patch of red* or yellow on the side of the head and yellow or olive vertical markings on the dark upper shell. Sliders prefer quiet water with plenty of vegetation, and rarely venture far on land. They are found across the South from Maryland to Texas and as far north as Illinois.

REDBELLY TURTLE **10–15 in.**

Has *reddish markings* on upper shell. Usually prefers larger lakes and rivers. Found from New Jersey to North Carolina, Florida, and in two small areas of Alabama and Massachusetts.

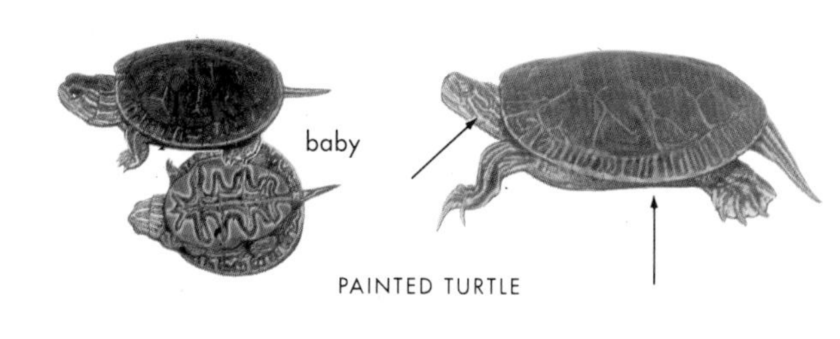

PAINTED TURTLE

RIVER COOTER

SLIDER

REDBELLY TURTLE

Map Turtles

These shy turtles of lakes and rivers are quick to plunge into the water if disturbed. The name comes from the intricate maplike markings on the shell. All have a ridgelike keel down the back, often with points like a saw blade. Females grow to be much larger than males.

COMMON MAP TURTLE 7–11 in. (female)

Yellow spot behind eye. Prefers large bodies of water, where it eats snails and crayfish. Slow to hibernate in the winter; can sometimes be seen swimming under ice. Found from Quebec to Alabama, west to Minnesota and Arkansas.

FALSE MAP TURTLE 5–11 in. (female)

Also has *yellow spot behind eye*, but keel is higher, with obvious knobs. Will often climb up slippery snags to reach basking spots where other turtles would never be seen. Found in Mississippi River Valley and nearby.

Box Turtles

Box Turtles are dry-land turtles with a hinged lower shell.The hinge allows them to close the shell so tightly a knife blade cannot be inserted—unless the turtle is too fat to fit completely into its shell. They are in the same family as the pond, map, and other water turtles described above, but they look a little like the tortoises on page 65.

EASTERN BOX TURTLE 4–7 in.

The high, domelike shell of the Eastern Box has yellow, orange, or olive markings on a dark background. It and its relatives were once widespread in the East, but are now rarely seen in some areas.

ORNATE BOX TURTLE 4–6 in.

A turtle of the plains and prairies, with both the top and bottom shell *elaborately marked.* Found from Indiana to Wyoming and south through Texas.

COMMON MAP TURTLE

FALSE MAP TURTLE

EASTERN BOX TURTLE

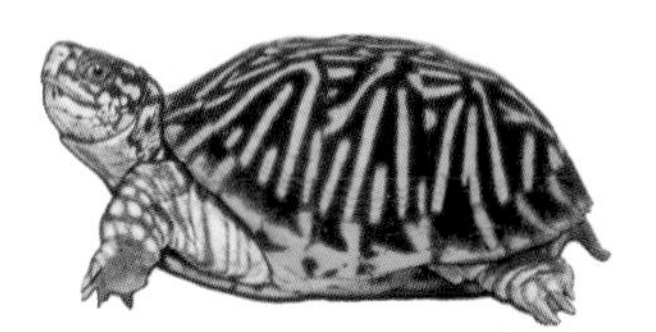

ORNATE BOX TURTLE

Tortoises

Unlike other turtles, tortoises live strictly on land. They are found in every continent except Australia. The feet are stumpy, especially in the rear, and are unwebbed. Most are vegetarians, but some eat meat occasionally.

GOPHER TORTOISE **6–15 in.**

Sometimes confused with the box turtle, but the Gopher has stumpy, *elephantlike feet*, and its lower shell is not hinged. It digs burrows up to 35 feet long, in which other animals, from owls to rattlesnakes, may seek shelter. In warm weather the Gopher Tortoise usually emerges from its burrow each morning to forage on grass, leaves, and fruit. Found in sandy parts of the coastal plain from South Carolina to Louisiana and southern Florida.

TEXAS TORTOISE **5–8 in.**

This tortoise's upper shell may be nearly as wide as it is long, and like the Gopher's it is tan to brown in color. Also has *stumpy feet*. Lives in drier lands than the Gopher Tortoise and often rests in a shallow scrape under a plant instead of digging a burrow. Prowls actively in hot weather, usually in early morning and late afternoon, searching for grass, prickly-pear cactus, and other vegetation to eat. Found in southern Texas and Mexico.

DESERT TORTOISE **8–15 in.**

Like the other tortoises, the Desert Tortoise has a domed shell and *stumpy legs*. Its snout is more rounded than the Texas Tortoise's, and the shell looks sculpted. Its burrows, 3 to 30 feet long, are often found at the base of bushes and may be occupied by one to many tortoises. Short tunnels are used for temporary shelter, longer dens for hibernation. A *threatened species*, the Desert Tortoise is found in the deserts of the Southwest, from Nevada through Arizona and California to Mexico.

GOPHER TORTOISE

TEXAS TORTOISE

DESERT TORTOISE

Sea Turtles

Large turtles with low, streamlined shells and powerful flippers, they live primarily in tropical seas but may be found much further north, especially in the Atlantic.

HAWKSBILL **30–36 in.**

The "*tortoiseshell*" turtle, endangered by being hunted for its shell. Generally brown, darker when young. Has a sharp, hawklike beak that it uses to defend itself. Hawksbills are found on both coasts. In the Pacific, they range from Peru to southern California, and in the Atlantic from southern Brazil to New England.

GREEN TURTLE **30–60 in.**

The Green Turtle is actually brown in color (sometimes olive). Its name comes from the greenish fat of its succulent flesh, which is prized as a food. It rarely comes ashore except to lay eggs at communal nesting sites, where it is easy prey for predators. In the Pacific it only rarely gets as far north as San Diego, but in the Atlantic it ranges to Massachusetts.

LEATHERBACK **48–96 in.**

The largest living turtle, weighing up to a ton. Its shell is not divided into plates like other turtles' but is covered with a smooth, *leathery, blackish skin* with prominent *ridges*. Feeds chiefly on jellyfish. A powerful swimmer, it ranges throughout the world, as far north as Alaska and Newfoundland. Its extraordinary ability to keep its deep-body temperature considerably higher than the surrounding water allows it to survive in colder regions than other turtles.

HAWKSBILL

GREEN TURTLE

LEATHERBACK

LIZARDS

Lizards look similar to salamanders, but unlike them, lizards have scales and claws.

Be careful when seizing any lizard, because the tails of many species break off at the slightest pinch. A new tail usually grows, but it often looks different from the old one.

Geckos

Geckos are often seen climbing up and down walls and even across ceilings in houses in the tropics. Most Geckos are nocturnal and often call at night. Their name is based on the call of a common Asian species.

WESTERN BANDED GECKO **4–5½ in.**
Banded Geckos have *movable eyelids*. This species has brown bands on a pink or yellow background. Young ones have bold *chocolate brown and yellow bands*, but these become mottled as the lizard gets older. These Geckos stay close to the ground and are not as acrobatic as some of their relatives. Although it looks delicate, this lizard can live in the driest parts of the desert because it is active at night, otherwise staying hidden in rock crevices or underground. Lives in the dry Southwest and Mexico.

TEXAS BANDED GECKO **4–5 in.**
Similar to the Western Banded Gecko, but its dark bands are broader. Lives mostly in southwestern Texas and northeastern Mexico.

MEDITERRANEAN GECKO **4–5 in.**
A pale, ghostly lizard with very large, lidless eyes. Not native to the U.S., but has been introduced into many areas in the southern half of the country and is now an "urbanized" Gecko that lives in or near houses. It has *large wartlike bumps* over most of its body, and is marked with both light and dark spots on a pale ground color. On warm evenings, look for it on buildings, window screens, or near lights that attract insects.

WESTERN BANDED GECKO

TEXAS BANDED GECKO

MEDITERRANEAN GECKO

Anoles

Most of the 250 species of Anoles live in the tropics. Only one is native to the mainland U.S.

GREEN ANOLE **5–8 in.**

Often mistakenly called a "chameleon." Its small size, plain green hue and *pink throat fan* make this lizard unique. It can change its color from green to brown, but its color-changing abilities are poor compared to the true chameleons of the Old World. Abundant in the South, where it may be seen on fences and in trees. Easy to catch at night with the aid of a flashlight.

Tree and Side-blotched Lizards

Both of these related lizards have a fold of skin across their throats, and neither has large, strongly keeled scales.

SIDE-BLOTCHED LIZARD **4–5½ in.**

This lizard has a *single bluish black spot* behind its armpit. Males have a back that is spangled with blue flecks. It is one of the most abundant lizards in the arid regions of the West, and eats scorpions, spiders, and insects.

TREE LIZARD **4–5½ in.**

A small, gray or grayish brown lizard that spends much of its time climbing. It has folds of skin along its body that give it a somewhat wrinkled look. It is usually nearly perfectly camouflaged in the trees or against rocks, and it frequently rests in a vertical position, often head down. When chased, it is adept at dodging, keeping on the opposite side of the tree trunk or rock. The Tree Lizard lives in the Southwest and Mexico.

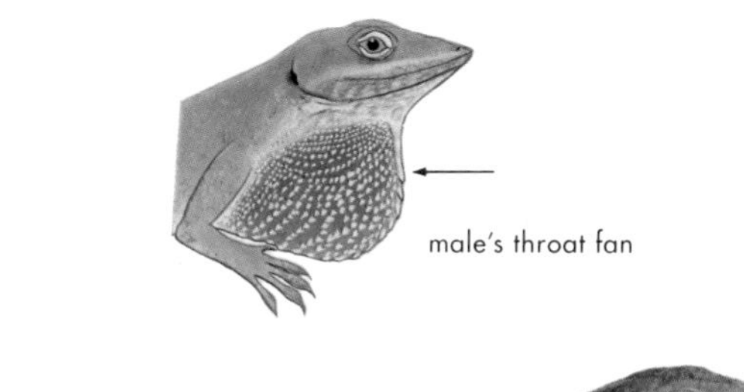

GREEN ANOLE

SIDE-BLOTCHED LIZARD

TREE LIZARD

Collared and Leopard Lizards

Despite their different head shapes—narrow in Leopard Lizards and almost grotesquely enlarged in Collared Lizards—the members of these two groups are closely related. All are alert, elusive, and pugnacious.

COLLARED LIZARD **8–14 in.**

This gangling, big-headed, long-tailed lizard can run on its hind legs like a miniature dinosaur. Look for the *two black collar stripes*. Its other coloration varies; may be greenish, brown, or yellowish. It is a mostly western reptile that prefers hilly, rocky areas. Collared Lizards may be found basking on rocks. If you catch one, handle it carefully to avoid being nipped.

LONGNOSE LEOPARD LIZARD **8–15 in.**

Here is a "leopard" that can change its spots, from black to very light brown. Ground color is gray, brown, or yellowish, but breeding females like the one shown have reddish markings. Lives in the West, mainly in dry, flat, sandy places.

Spiny Lizards

These lizards have ridges (keels) on the scales of their backs. Some are so rough they seem almost like pine cones with legs and tails.

DESERT SPINY LIZARD **7½–13 in.**

A stocky, light-colored lizard with large, pointed scales and a *black wedge-shaped mark* on each side of its neck. Lives in Western deserts and dry grasslands. It is a good climber, but hides under objects on the ground. Often bites when captured.

CREVICE SPINY LIZARD **5–11½ in.**

The *banded tail* and *broad black collar* can be seen from a considerable distance. The male has a bright blue throat. Boulders and rocky places in the West are favorite habitats. When chased, it wedges itself into a rock crevice so tightly it is difficult to pry it out unharmed.

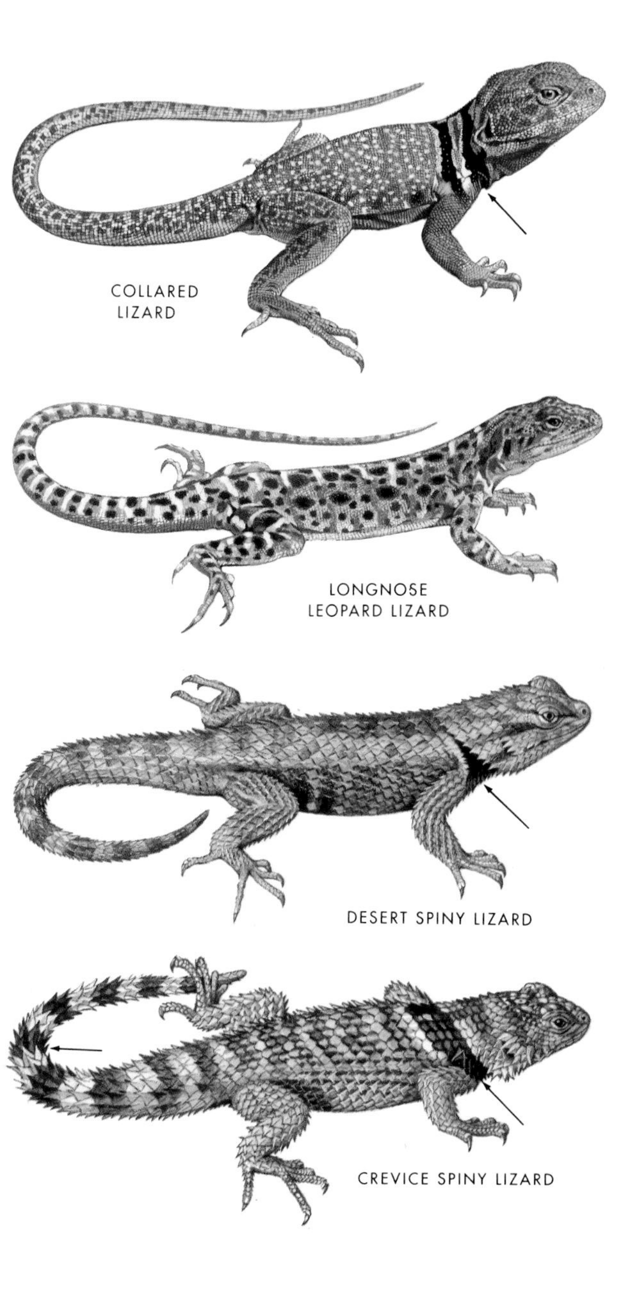
COLLARED
LIZARD
LONGNOSE
LEOPARD LIZARD
DESERT SPINY LIZARD
CREVICE SPINY LIZARD

WESTERN FENCE LIZARD **4–8 in.**

One of the most common western lizards, also known as "blue-belly" and often seen on rocks, fenceposts, and the sides of buildings. It is black, gray or brown with a blotched pattern and *blue markings* on the belly. Male usually also has a blue throat. The Western Fence Lizard occupies a great variety of habitats, from farmland to grassland to forest, but it avoids the driest parts of the desert. Its range extends from Washington to Baja California.

EASTERN FENCE LIZARD **4–7 in.**

A small gray or brown spiny lizard that lives in forested areas. Females have a series of *dark wavy lines* across their backs; males have *blue throats.* The Eastern Fence Lizard is the only spiny lizard in most of its range, which covers much of the eastern and central U.S. It is often seen on rail fences or on rotting logs and stumps. When surprised on the ground in wooded areas, a Fence Lizard will usually dash for a nearby tree, climb upward a short distance, and then "freeze" on the opposite side of the trunk. Western varieties, with other types of markings, live on the plains and prairies, and seldom leave the ground.

SAGEBRUSH LIZARD **4–6 in.**

Resembles the Western Fence Lizard, including blue belly markings, but the Sagebrush Lizard is smaller. It often has a black bar on its shoulder. As its name implies, it is common in sagebrush habitats but also lives in other brushlands, forests, and along river bottoms in the coastal redwood forests. Its range extends over much of the western U.S. Eats insects, mites, ticks, spiders, and scorpions.

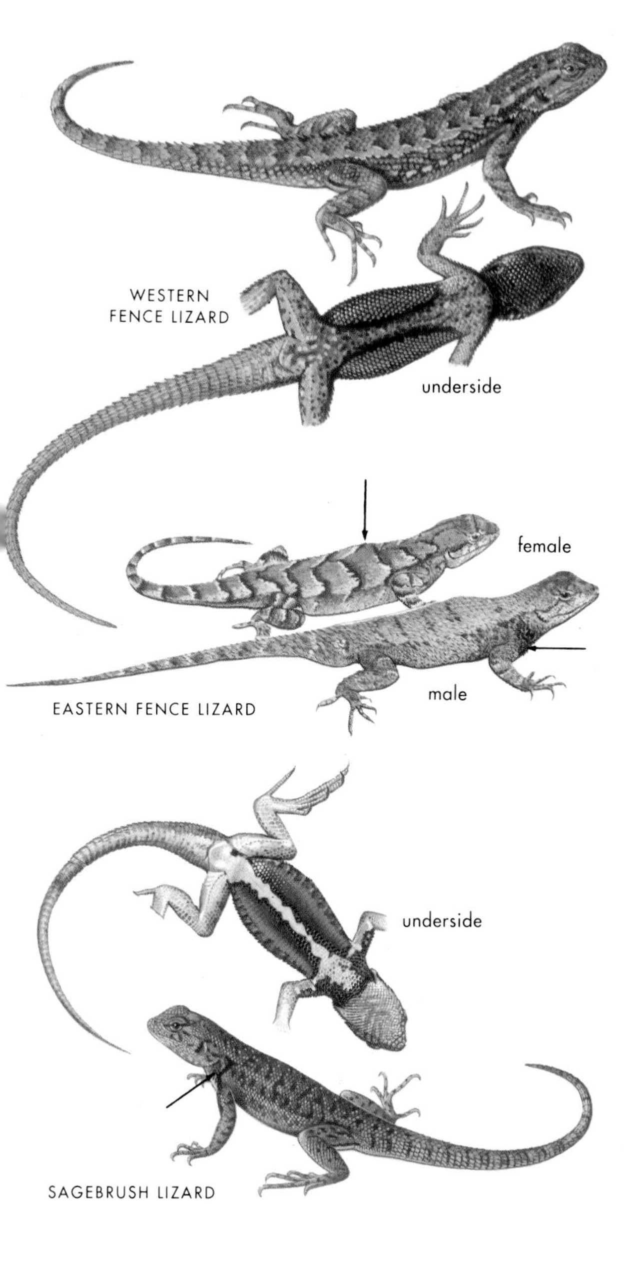
WESTERN
FENCE LIZARD
underside
female
male
EASTERN FENCE LIZARD
underside
SAGEBRUSH LIZARD

Some Non-spiny Western Lizards

FRINGE-TOED LIZARDS **6–9 in.**
Flattened lizards with velvety skin, toes fringed with large scales that help prevent it from sinking into the sand, and large earflaps—all adaptations to life in the sand. Their sandy ground color and black flecks are good camouflage in the loose sand of the dunes. When running at top speed, Fringe-toed Lizards run on their hind legs. If chased, they may dart behind a bush or burrow quickly into the sand. This lizard lives in the deserts of the Southwest. There are three closely related species.

DESERT IGUANA **8–15 in.**
A large, pale, round-bodied lizard with a long tail and a rather small, rounded head. It has a *single row of pointed scales* down its back. Lives in the desert Southwest, usually in areas where creosote bushes, one of its main foods, grow. It can withstand great heat, staying out in the sun when most other lizards take shelter.

BANDED ROCK LIZARD **6–10 in.**
This flat-bodied lizard has a *single black collar* and a *banded tail.* Its ground color is gray, brown, or olive, with many small white or bluish spots. It is always found in rocky places, around which it crawls easily, with its legs held well away from its body, its body low, and its hindquarters swinging from side to side. It prefers massive rocks in the shady, narrow parts of canyons. The small range of the Banded Rock Lizard barely extends north into California from Baja California.

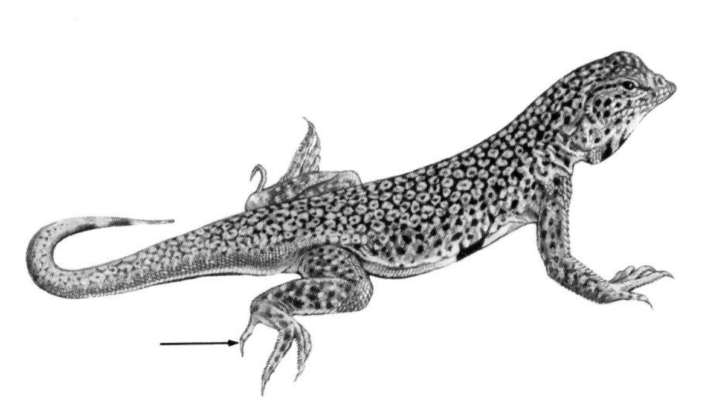

FRINGE-TOED LIZARD

DESERT IGUANA

BANDED ROCK LIZARD

CHUCKWALLA **10–16 in.**

This large, flat, dark lizard has *loose folds of skin* on its neck and sides. Males are usually light gray or reddish with black heads and legs; females and young lizards may have dark crossbars. Chuckwallas are often seen sprawled on rocks in the sun. They are quite common in the desert Southwest, where nearly every rocky hillside and outcrop will have its Chuckwalla. To find one, watch for basking individuals on rocks and take note of which crevice the lizard enters, or listen for the sandpapery sound as it slides into a crack. "Chucks" are hard to capture, however, for once they get into a crevice, they gulp air, puffing up and wedging themselves in tightly.

GILA MONSTER **11–20 in.**

This is the only venomous lizard in the U.S. It has a large, heavy body, a short swollen tail, and a gaudy pattern of black with pink, orange, or yellow. Its coloring helps hide it in dim light; the light markings look like sticks or rocks on a dark background. The dark tongue flicks out like a snake's. It prefers dry shrubby and grassy areas and deserts, ranging from extreme southwestern Utah, southern Nevada, southeastern California, and Arizona into Mexico. Takes shelter in the burrows of other animals and under rocks. The Gila Monster eats small mammals as well as other reptiles. It stores fat in its tail. Approach this lizard with care! Its bite is tenacious and extremely painful, though it is seldom fatal to humans. Gila Monsters are not dangerous unless they are molested, and they should not be killed.

CHUCKWALLA

GILA MONSTER

Horned Lizards

These are the misnamed "horny toads," most of them adorned like a cactus. When handled, some have the alarming habit of sometimes squirting blood from the corners of their eyes for a distance of several feet. In many states, horned lizards are protected by law and may not be removed from their homes.

TEXAS HORNED LIZARD **2½–7 in.**
Dark stripes radiating from the eyes identify this lizard. Its color usually matches the soil around it. The two central horns are quite long and sharp. Lives in open country where there is loose soil in which it can bury itself, in Texas and nearby states. This lizard is often carried home—illegally—by visitors, but it does not live very long in captivity. This may be because it eats great quantities of live ants, which its keepers are hard-pressed to supply.

SHORT-HORNED LIZARD **2½–6 in.**
Has *short, stubby horns.* Generally gray or tan with brown blotches. This lizard has a large western range, from Canada to Mexico, and is found from the plains into the mountains. This species is more tolerant of cold than other Horned Lizards.

COAST HORNED LIZARD **2½–5½ in.**
This Horned Lizard has *two rows of pointed scales* on each side of its body. It is colored much like its cousins and has large, wavy, dark blotches. It needs open, sunny areas and patches of loose soil where it can bury itself. Lives in most of California west of the desert, and throughout Baja California.

DESERT HORNED LIZARD **2½–5½ in.**
The desert counterpart of the Coast Horned Lizard. This lizard has a very blunt snout, and its horns and body spines are short. *One row of pointed scales* on the side of its body. Its general color matches the surrounding soil. Its range extends from Idaho to Baja California.

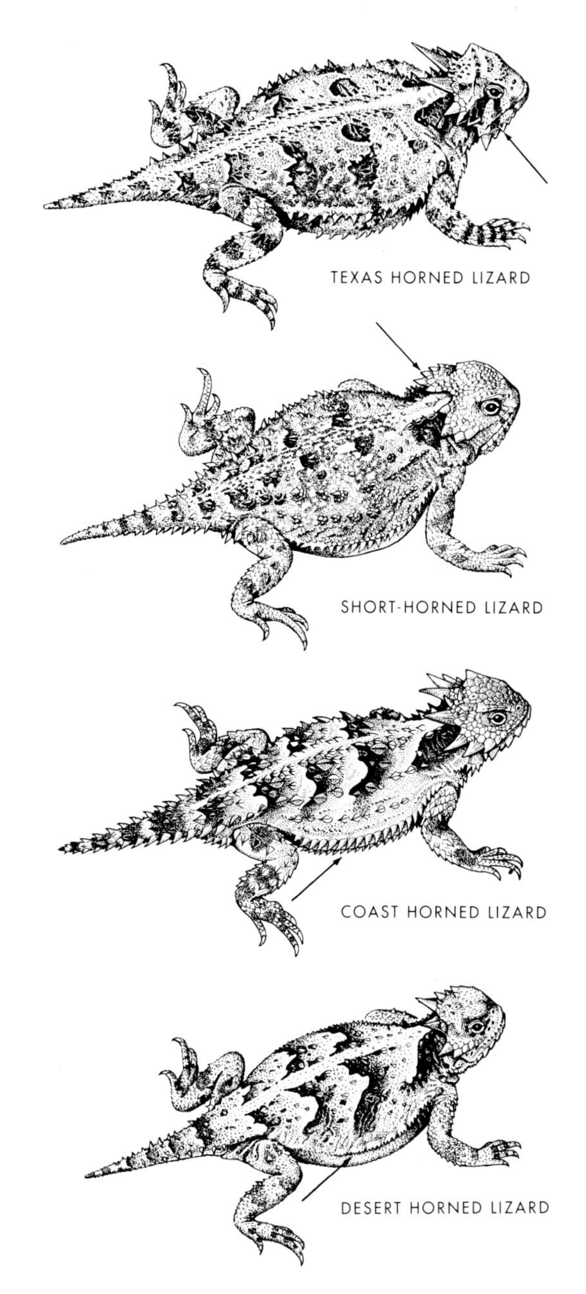
TEXAS HORNED LIZARD
SHORT-HORNED LIZARD
COAST HORNED LIZARD
DESERT HORNED LIZARD

Alligator Lizards

These lizards have short legs and extremely long tails that may account for more than half their total lengths. Their scales are hard and bony, which makes their bodies rather stiff and rigid. A strip of soft scales on their sides allows the body to expand for breathing and to make room for food.

SOUTHERN ALLIGATOR LIZARD **7½–15 in.**

Usually has dark crossbands on its back and tail, on a ground color of reddish or brown. Eyes are pale yellow. Like other Alligator Lizards, its long tail snaps off so easily that full-tailed individuals are uncommon. When the tail grows back, it is shorter. This lizard's range is near the West Coast. It lives in grasslands and woodlands, and may be found in woodpiles and trash heaps near houses. Eats insects, scorpions, and spiders, including black widows.

NORTHERN ALLIGATOR LIZARD **6½–14 in.**

This lizard may have crossbands, but they are not very regular, and some individuals are plain gray, greenish, or bluish. Eyes are usually dark. It prefers cooler and damper places than those favored by the Southern Alligator Lizard, and is often found inside rotten logs and under rocks. Lives chiefly in woodlands in the Pacific Northwest, from California north into Canada.

MADREAN ALLIGATOR LIZARD **6–15 in.**

Also called the Arizona Alligator Lizard. Distinct wavy crossbars mark this lizard, standing out against a pale background of gray, tan, or brown. Young lizards also have contrasting dark crossbars. Its eyes are orange or pink. The Madrean Alligator Lizard lives mainly in the mountains, often in rocky places near streams. Its range includes parts of Arizona, New Mexico, and Mexico.

SOUTHERN ALLIGATOR LIZARD

NORTHERN ALLIGATOR LIZARD

MADREAN ALLIGATOR LIZARD

Earless and Zebratail Lizards

With their slim bodies, flat tails, and long, slender legs, these lizards are well suited for running quickly.

LESSER EARLESS LIZARD **4–5 in.**

A small, ground-living lizard with smooth skin and no ear openings. Its color may be brown, gray, or whitish, often matching the surrounding soil. Usually has scattered light spots on its back and a pair of black marks on each side of its belly. The female develops a vivid orange or yellow patch on her throat during the breeding season. Lesser Earless Lizards live in the Great Plains and the Southwest. They are not particularly fast runners and can sometimes be caught by hand. When chased, Earless Lizards may dive into the sand and bury themselves with a shimmying motion.

GREATER EARLESS LIZARD **3–8 in.**

This lizard has no ear openings, and there are *black bars* on the underside of its tail. The bars are flashed as the lizard curls its tail over its back when it runs, waving from side to side as it slows to a halt. This may serve to distract predators to the tail, which can be sacrificed. Two black crescents mark the belly. Ground color is grayish for those living on gray soil, reddish for those living on red soil, etc. The value of this camouflage is seen when the lizard dashes a short distance away, settles down on sand or a rock, and virtually disappears. Native to Texas and the Southwest.

ZEBRATAIL LIZARD **5–8 1/2 in.**

Resembles the Greater Earless Lizard but has ear openings, and black belly markings are farther forward. The greyhound among lizards, reaching speeds of 18 miles per hour. Curls its tail, displaying the stripes, when it runs. Lives in the West, from Nevada south into Mexico, preferring hard, dry, open areas where there is room to run.

LESSER EARLESS LIZARD

GREATER EARLESS LIZARD

ZEBRATAIL LIZARD

Whiptails

These lizards can be recognized by their long tails and by their active, nervous prowling. Some species of Whiptails are unisexual, or "all-female." Adults lay eggs, which develop unfertilized and hatch into females. Males are rare in these species.

CHECKERED WHIPTAIL **8–15½ in.**
One of our largest Whiptails. Color varies—some are checkered, others have rows of black spots. Ground color is yellowish to cream. Lives mainly in Texas and New Mexico. It uses many habitats—plains, canyons, foothills, and river floodplains (as along the Rio Grande)—but there are almost always rocks nearby. Checkered Whiptails are unisexual.

WESTERN WHIPTAIL **7½–15 in.**
Back and sides with spots, bars, or a network of black markings on a paler background. Young ones have a bright blue tail. An active lizard of deserts and other dry habitats, especially where plants are sparse and there is room to run. Also found in open woodlands, but it avoids dense grasslands and brushy areas. Native to much of the western U.S. and northern Mexico. There are two sexes.

RACERUNNER **6–10½ in.**
A small, unspotted Whiptail with *6 or 7 light stripes*. Head and body color between the stripes is greenish in the West, dark in the East. This species comes in both sexes, and males have a blue belly. Young ones have a light blue tail and bright yellow stripes. The Racerunner is an active lizard, conspicuous because of its boldness. It likes open areas covered with sand or loose soil, such as fields, rocky outcrops, and river floodplains. Racerunners are well named, usually winning any race with a would-be collector.

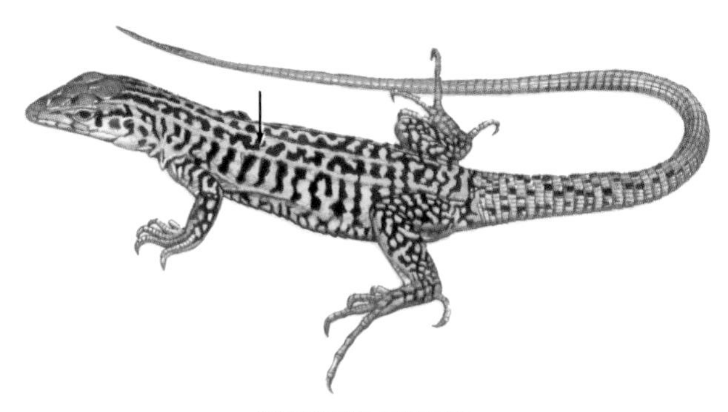

CHECKERED WHIPTAIL

WESTERN WHIPTAIL

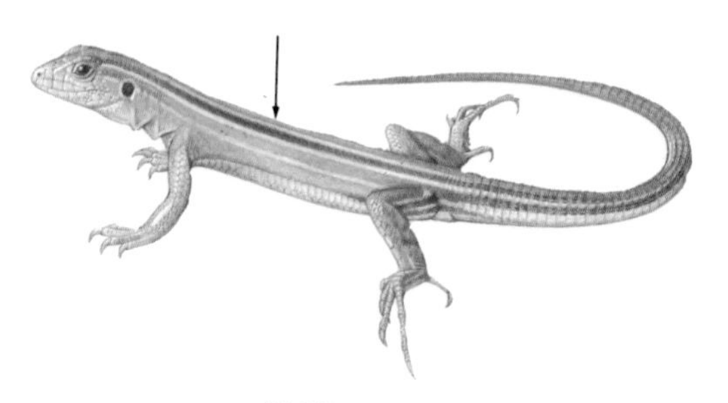

RACERUNNER

Skinks

These smooth, shiny, wary, and active lizards are hard to catch and hard to hold. Our species have smooth, round-edged scales. Watch the tails—they break off easily. Most Skinks will try to bite.

GROUND SKINK **3–6 in.**

This elfin reptile of the woodland floor quietly but nervously searches for insects among leaves and decaying wood. It is a small, smooth, golden brown to blackish brown lizard with a *dark stripe* down each side. It has a "window" in its lower eyelid that lets it see when its eyes are closed. When running, it makes side-to-side movements like a snake. The Ground Skink is likely to be found almost anywhere in the Deep South, even in towns and gardens.

COAL SKINK **5–7 in.**

Look for the *four light stripes* that extend onto the tail. When chased, the Coal Skink (as well as several other kinds of Skinks) does not hesitate to take refuge in shallow water, going to the bottom and hiding under stones. Its favorite habitats are humid, wooded hillsides, springs, and rocky bluffs overlooking creek valleys. Lives in scattered areas from the Great Lakes southwest to Texas and the Gulf Coast.

FIVE-LINED SKINK **5–8½ in.**

The color and markings of this Skink vary depending on its age and sex. Very young ones have *five white stripes* on a black ground color, and their *tails are bright blue.* As they grow older, the stripes fade and the ground color lightens, and the tail turns gray. Males have *orange-red jaws* during breeding season. Females usually keep their stripes. Over most of its range, which includes most of the eastern U.S., this Skink is terrestrial, living in damp areas such as cutover woodlots and woodpiles. In Texas, the Five-lined Skink usually is found in trees.

GROUND SKINK

COAL SKINK

FIVE-LINED SKINK

GREAT PLAINS SKINK **$6^{1}/_{2}$–14 in.**

Our largest skink, and unique because the scale rows on its sides run on a *slant* instead of lengthwise. It is light gray or brown, with many dark spots that unite in places to form stripes. Young ones are black, with a blue tail and *orange and white spots* on their heads. This Skink is chiefly a grassland animal of the Great Plains, where it prefers fine soil for burrowing and sunken rocks for shelter. It is secretive and nervous, seldom seen except when rock slabs or other hiding places are overturned. Can inflict a painful bite.

WESTERN SKINK **5–8 in.**

Has a broad brown stripe down its back that is edged with black and bordered in turn by two whitish stripes, one on each side. The tail is blue, gray, or rusty. Young ones have a bright blue tail, and their stripes are quite vivid. The Western Skink prefers rocky habitats near streams where there is plenty of plant life to hide under, but it is also found on dry hillsides far from water. It is active in the daytime but usually stays out of sight. Found in western North America, from Canada south through Utah and down to the tip of Baja California.

MOLE SKINK **$3^{1}/_{2}$–$6^{1}/_{2}$ in.**

Mole Skinks are so slender and their legs are so short that they look and act much like snakes. They usually have a reddish tail, like the one shown, but it may also be blue, brown, or orange. Mole Skinks like to burrow in the sand. They hide in debris along the shore, in sand hills with some plant cover, or in dry, rocky areas. They are found chiefly in Florida, but also in parts of Georgia and Alabama.

GREAT PLAINS SKINK

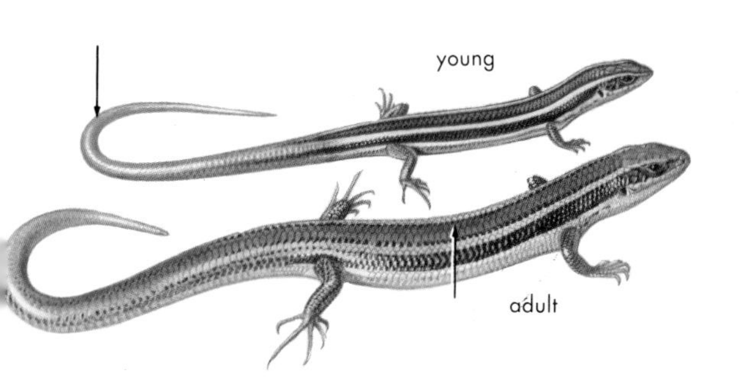

WESTERN SKINK

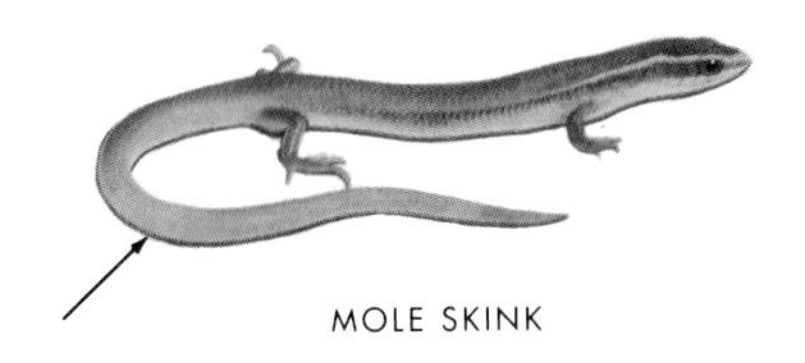

MOLE SKINK

Legless Lizards

These lizards look like snakes, but unlike snakes they have movable eyelids.

CALIFORNIA LEGLESS LIZARD 7–11 in.
A burrowing reptile about the size of a pencil. Although its eyes are small, you should be able to see the eyelids blink in good light. It is usually silver or tan above, yellow below, and the skin looks polished. Legless Lizards live in habitats where there is loose sand or soil to burrow in, moisture, warmth, and plant cover. When picked up, a Legless Lizard may probe your hand with its snout with surprising force. Found mostly near the coast of California and Baja California.

Glass Lizards

These lizards are easily mistaken for snakes, but look for the movable eyelids. All have long tails that break off easily in most species. Like Alligator Lizards, Glass Lizards have rigid, bony scales, compensated for by a flexible groove down the sides of their body to allow the body to expand for breathing and food.

SLENDER GLASS LIZARD 22–42 in.
Usually has a *dark stripe* down the middle of its back, plus several *narrower dark stripes* on the lower half of its body. Old adults may be brown. The Slender Glass Lizard is usually found in dry grasslands or dry, open woodlands. It seldom burrows, except when it hibernates. When captured, it tries vigorously to escape, lashing back and forth, and may snap off its own tail. Found in the central and southeastern U.S.

EASTERN GLASS LIZARD 18–42 in.
Often a plain greenish color, with no stripes. Most often found in wet meadows and grasslands and pine woods. The tails of Glass Lizards are so fragile that full-tailed specimens are not common. The regenerated tip may be so sharply pointed that some people think it is a stinger. Lives in the Southeast, from Virginia to Louisiana and throughout Florida.

CALIFORNIA LEGLESS LIZARD

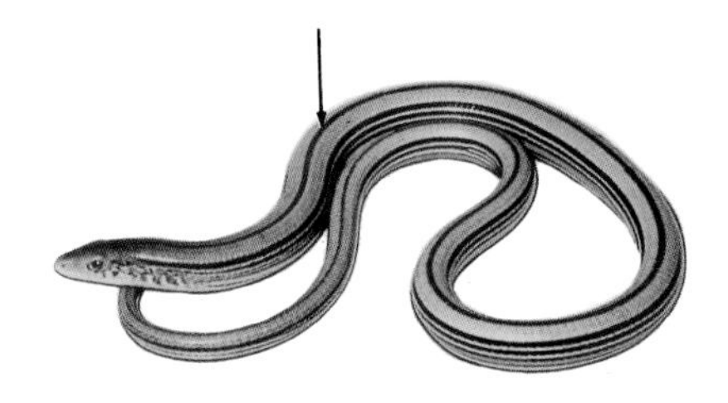

SLENDER GLASS LIZARD

EASTERN GLASS LIZARD

SNAKES

Snakes are closely related to lizards, but all are legless and have no eyelids. Venomous snakebite is rare in most parts of the U.S., and seldom fatal. Still, you should be able to identify all venomous snakes in your area. If you are bitten by a dangerous snake, get medical help at once. The best "snakebite kit" consists of car keys and some coins with which to call a hospital to say you are coming.

Blind Snakes

These snakes are sometimes called "worm snakes" because they resemble earthworms.

WESTERN BLIND SNAKE **7–16 in.**
This slim snake has a blunt head and tail and a silvery sheen. Its eyes are no more than *dark spots,* but it has little need for eyes since its prey, mainly ants and termites, lives underground. It occurs in the Southwest and Mexico, where its habitats include beaches and rocky hillsides.

Boas

Boas are thick-bodied snakes that kill their prey by squeezing it. Ours are dwarfed compared with their relatives, such as the Boa Constrictor. When alarmed, our boas may roll into a ball and hide their heads among the coils.

RUBBER BOA **14–33 in.**
This stout snake looks and feels like rubber. Sometimes called the "two-headed snake" because the *tail is shaped like the head.* Plain brown or green above. Lives in grasslands, woodlands, and brush in the northwestern states and California.

ROSY BOA **22–44 in.**
Rosy, slate, or tan above, with *three wide brown or gray stripes* or irregular brown patches. It lives in rocky shrublands and deserts in the Southwest, where it is attracted to oases and streams. Eats small mammals and birds.

WESTERN BLIND SNAKE

RUBBER BOA

ROSY BOA

Colubrids: Water Snakes and Their Allies

Three quarters of all the world's snakes belong to the enormous Family Colubridae. Included in the large subgroup on pp. 96–103 are the typical Water Snakes, the Garter and Ribbon Snakes, and many smaller kinds, such as the Brown Snake. All usually are found close to water or in damp environments.

DIAMONDBACK WATER SNAKE 30–63 in.

The light areas on the back are vaguely diamond-shaped, but they are better described as dark brown *chainlike* markings on a ground of lighter brown or dirty yellow. This is a common snake in most of its range, from the south-central U.S. to Mexico, appearing in all kinds of habitats, from along rivers to cattle tanks.

NORTHERN WATER SNAKE 22–59 in.

These harmless snakes are often seen basking on logs, branches, or brush from which they drop or glide into the water at the slightest alarm. The only large water snake in most of the northern states comes in a bewildering variety of forms. Its ground color varies from pale gray to dark brown, and its markings, which may be blotches, crossbands, or bars, vary from bright reddish brown to black. The Northern Water Snake ranges from Maine to Colorado and south to the Gulf of Mexico. It can be found in virtually every unpolluted swamp, marsh, bog, stream, pond, or lake in its range.

SOUTHERN WATER SNAKE 22–62 in.

This snake has dark crossbands, often black-bordered, and a dark stripe from its eye to its jaw. The bands may be yellow, red, brown, or black; the ground color could be gray, tan, yellow, or reddish. Some individuals darken with age, even becoming nearly black. The Southern Water Snake occupies freshwater habitats, such as streams, ponds, and lakes. It lives in the southeastern U.S., from North Carolina to Texas and all of Florida.

DIAMONDBACK WATER SNAKE

NORTHERN WATER SNAKE

SOUTHERN WATER SNAKE

PLAINBELLY WATER SNAKE 30–62 in.
A snake that has many different colors and patterns, depending on where you find it. In the Southeast, it most likely is a brown snake with a red belly. Another form has a yellow belly. Farther west, in and around Texas, it usually has large blotches, like the young one shown at right. All forms live in or near river bottoms, swamps, or lakes, but they often wander well away from water in hot, humid weather.

COMMON GARTER SNAKE 16–51 in.
Like the fancy garters that once were fashionable for holding up men's socks, these snakes have longitudinal stripes, one on each side of the body. Most also have a pale yellow or orange stripe down their backs. In one form or another, this snake is found in nearly all of North America except parts of the arid Southwest. It is extremely variable in its color and markings, though the *three yellowish stripes* are usually well defined. A beautiful red form called the San Francisco Garter Snake is *endangered*. You will find the Common Garter Snake in a variety of environments, including ponds, woods, farms, and city lots. It is a spirited snake that will defend itself energetically when cornered.

CHECKERED GARTER SNAKE 18–42 in.
A Garter Snake with a *checkerboard pattern*. Black, squarish spots often overlap the light stripes. This snake lives in Texas and the Southwest, where it stays close by streambeds, springs, irrigation ditches, or other places where water may be present. Its menu includes fish, toads and frogs, tadpoles, and lizards.

PLAINBELLY WATER SNAKE

COMMON GARTER SNAKE

CHECKERED GARTER SNAKE

PLAINS GARTER SNAKE **15–43 in.**

This Garter Snake has a well-defined orange or yellow stripe down its back and black bars on its lips. Garter Snakes bear live young, and the Plains Garter Snake sometimes gives birth to 90 or more babies at once, each about 7 inches long! It is an abundant snake in most of its range, especially common in river valleys and near prairie ponds. It used to be common in city lots and parks, but building and the use of pesticides has made it much rarer in those habitats.

WESTERN TERRESTRIAL GARTER SNAKE **18–43 in.**

This snake's colors vary with the individual, but the *yellowish stripe* down its back is usually easy to see, and it also has two side stripes. The pale ground color may be checkered with dark spots, or it may be dark with white flecks. It is called "terrestrial" because it is usually found on dry land, but it sometimes enters water. Lives in the western U.S. and Canada, from sea level to the mountains.

RIBBON SNAKES **18–38 in.**

Ribbon Snakes are very slim and trim. The *three pale stripes* are well set off against the dark slender body, and they have unmarked "lips." Our two species are highly adaptable and live in a variety of habitats, from dry grasslands to the tropics. They favor the plant life around streams, lakes, and marshes. When frightened, they often take to the water and swim with speed and grace. They are good climbers, and are often first seen as they drop into the water from sunbasking in the trees. Their combined ranges extend from New England to southern Florida and New Mexico southward to Central America.

PLAINS GARTER SNAKE

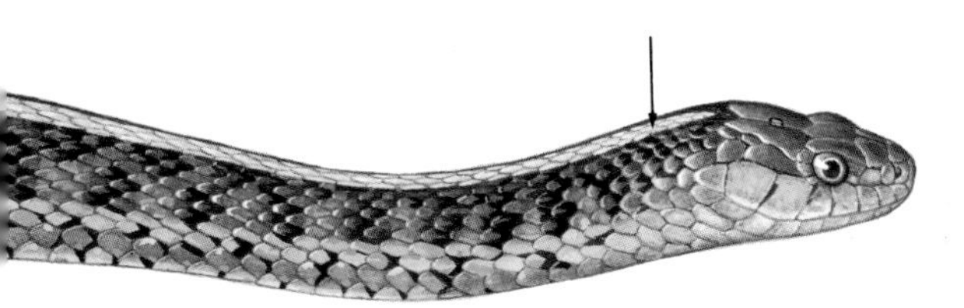

WESTERN TERRESTRIAL
GARTER SNAKE

RIBBON SNAKE

BROWN SNAKE **9–21 in.**

This is a small, secretive snake, brown and usually with two parallel rows of blackish spots down its back. Young ones have a conspicuous *yellowish collar* and are generally darker than adults. In the days before pollution and heavy pesticide use, Brown Snakes frequently turned up in parks, cemeteries, empty lots, and even large cities. Although they are still common in some areas, and may live close to people, they are such good hiders that few people are familiar with them. Away from cities, habitats include bogs, marshes, and moist hillsides. Like its relatives the Water and Garter Snakes, the Brown Snake may flatten its body when it is alarmed. Ranges widely over the eastern and central U.S. and south into Mexico.

REDBELLY SNAKE **8–16 in.**

Look for the *plain red belly* and the *three pale spots* on the neck. Otherwise, this snake may be brown, gray, or black. It is a shy snake that is particularly abundant in mountainous parts of the Northeast. Also lives in open woodlands and sphagnum bogs.

SMOOTH EARTH SNAKE **7–15 in.**

A small, earthy-colored (gray or reddish brown) snake with virtually no markings. Its belly is plain white or yellowish. It is skilled at staying out of sight, and is rarely seen above ground except after cool, heavy rains. Its range includes much of the eastern half of the U.S., from New Jersey to Texas. Earth Snakes eat earthworms and soft-bodied insects. Their habitats include abandoned fields, the vicinity of back roads and trails, and deciduous forests.

SMOOTH EARTH SNAKE

Colubrids: Rear-fanged Snakes

Several snakes have fangs located in the rear upper part of the mouth. Some are venomous, including the dangerous Boomslang and Twig Snake of Africa, whose bites can be fatal to humans. The Lyre Snake is our largest venomous rear-fanged snake. Its venom is considered to be weak and thus scarcely dangerous to humans. Still, it should be avoided or handled with care.

NIGHT SNAKE **12–26 in.**

This pale snake, marked with dark spots, often has a *pair of large dark blotches* on its neck. It has catlike vertical pupils and prowls at dusk and at night, looking for lizards, small snakes, and other prey, which it subdues by injecting venom with the large teeth in the back of its upper jaw. Ranges over large parts of the West.

LYRE SNAKE **18–48 in.**

A "cat-eyed" snake with vertical pupils. It is named for the dark *lyre-* or *V-shaped mark* on its head. The Lyre Snake is light brown or gray with brown blotches. Lives in the Southwest, in rocky areas where it can hide by day, emerging at night to hunt. It eats lizards, birds, and small mammals, including bats, which it catches at their roosts and stuns with an injection of venom.

SOUTHEASTERN CROWNED SNAKE **8–13 in.**

This snake has a *black cap* and a *black collar.* It is otherwise plain light brown or reddish brown, with a white belly. This snake is found in many kinds of habitats—it is where you find it, in swamps or dry wooded hillsides, wilderness areas and backyards. Look under things, especially stones and rotting logs. As its name implies, it lives in the Southeast, from Virginia to Louisiana.

NIGHT SNAKE

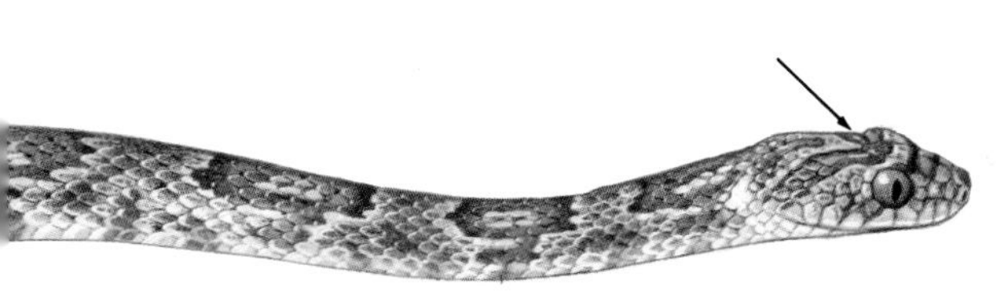

LYRE SNAKE

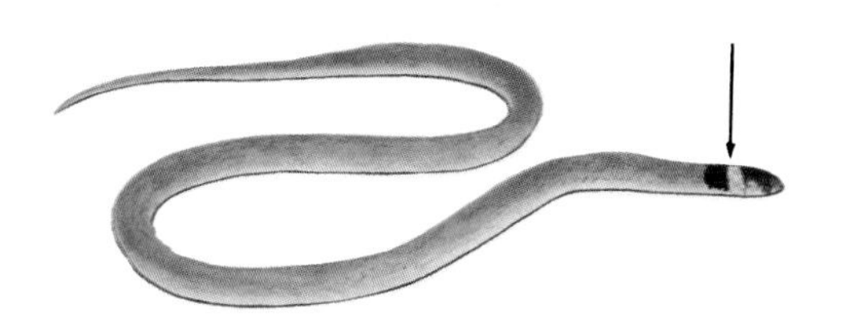

SOUTHEASTERN CROWNED SNAKE

Some Typical Colubrids

EASTERN HOGNOSE SNAKE 20–45 in.
Hognose Snakes have earned a bad reputation because of an alarming display of behavior in which they flatten their heads and necks, hiss loudly, and inflate their bodies with air. If this extraordinary show of hostility fails to deter an intruder, the snake will roll on its back, writhe convulsively for a few moments, then "play dead" with its mouth open and its tongue lolling. Turn the snake right side up, and it promptly rolls over again, giving the bluff away. This display, along with the *upturned snout,* are usually enough to identify this snake. Lives in the eastern half of the U.S., usually in sandy areas.

WESTERN HOGNOSE SNAKE 15–39 in.
Look for the *upturned snout.* This snake uses its snout for digging and its enlarged teeth for holding and perhaps deflating toads, its staple food. The habit of spreading its head and neck is not as highly developed as in the Eastern Hognose Snake, but it plays possum about as well. Some Western Hognose Snakes may crawl away without performing or hide their heads beneath their coils. Lives chiefly in sandy parts of the central U.S. and Mexico.

RINGNECK SNAKE 10–33½ in.
Usually a slender olive-green, bluish, brownish, or dark-colored snake with an orange or *golden collar* (collar missing in some parts of the Southwest). Its belly is yellowish to red in the West, yellow in the East. In the South, the underside of the tail may be red. When alarmed, the red-tailed Ringnecks twist their tails into coils and raise them, revealing the bright red color, a habit that has earned them the nickname "corkscrew snake." This secretive snake ranges over a large part of the U.S. It is most often found near streams and moist woods where there are plenty of places to hide.

EASTERN HOGNOSE SNAKE

WESTERN HOGNOSE SNAKE

RINGNECK SNAKE

SHARPTAIL SNAKE **8–18 in.**

This snake of California and the Pacific Northwest has a *sharp spine* at the tip of its tail. It also has long teeth, which it uses to eat slugs, its main food. Colored reddish brown or gray above, with bands of black and cream on its belly. It is a shy snake that lives in moist places, spending most of its time underground.

MUD SNAKE **40–81 in.**

A shiny, iridescent snake with a black back and a red belly, with the red color extending up the sides. This snake of the southern swamps and lowlands is a burrower, but also is thoroughly at home in the water. In fact, captives should be kept in aquariums. Feeds chiefly on eel-like salamanders. The young Mud Snake has a sharp tail tip, and when it is first caught, it will stab at the collector's hands with its tail. For this the Mud Snake has been called the "stinging snake."

RAINBOW SNAKE **27–66 in.**

An iridescent, glossy snake with red and black stripes. Before shedding, the skin turns a translucent blue that obscures the normal striped pattern. This southern snake is usually found in or near water. Streams passing through cypress swamps are a favorite habitat. It swims well and is adept at catching eels, its principal food. It may burrow in sandy fields near water and is sometimes turned up by plowing. The Rainbow Snake is usually inoffensive when handled, but when first caught it may thrash the hind part of its body about and stab its harmless tail at its captor's hands.

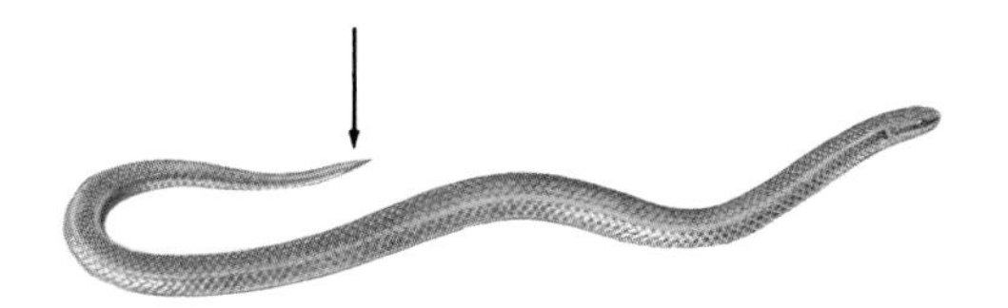

SHARPTAIL SNAKE

chin and neck

MUD SNAKE

chin and neck

RAINBOW SNAKE

RACER **20–73 in.**

This snake is found in many parts of the U.S., and its looks vary depending on where you find it. In the West, it is usually less than 36 inches long, plain brown or olive green above. In the East, it is often longer and usually black. There are also speckled and bluish forms. Adult snakes are usually unmarked, whereas young ones are blotched. This alert, active serpent is quick to flee but fights fiercely when cornered. If held by the neck with body dangling, it lashes its body back and forth, trying to break free. This snake throws a loop of its body over its prey, pinning it to the ground, and then swallows it whole.

COACHWHIP **36–102 in.**

A very large snake that lives in the southern half of the U.S. and Mexico. Its color and pattern vary widely with locality. In the East, there is a marked color change from black "forward" to light brown "aft." In the West, the Coachwhip may be reddish, yellow-brown, or dark brown with dark neck bands. There are also black and banded forms. Coachwhips use many kinds of habitats, chiefly prairies and deserts in the West and dry woods or creek valleys in the East. This active, fast-moving serpent sometimes prowls with its head raised high above the ground. It usually escapes a would-be collector with a burst of speed. Coachwhips make nervous captives, striking repeatedly at those who pass their cages.

SMOOTH GREEN SNAKE **12–26 in.**

A gentle little reptile with smooth scales that is bright green with a white or yellow belly. It is well camouflaged in plant growth, and lives in grassy and forested areas. Stays close to the ground, eating insects and spiders. Most common in the Northeast, but has a wide if spotty range that extends to Utah and New Mexico. The similar Rough Green Snake of the Southeast is longer and has keeled (ridged) scales.

RACER

two variations

COACHWHIP

SMOOTH GREEN SNAKE

INDIGO SNAKE **60–103 in.**

The Indigo Snake is the heaviest of all our non-venomous snakes. It is entirely shiny bluish black, except for some reddish brown color on its chin and the sides of its head. When cornered, the Indigo Snake flattens its neck vertically, hisses, and vibrates its tail, producing a rattling sound. It seldom tries to bite. Prey includes small mammals and other snakes, even Cottonmouths and Rattlesnakes. One race lives in the Southeast, primarily Florida, and another lives in Texas and Mexico.

CORN SNAKE **24–72 in.**

A long, slender snake that looks very different in various parts of its wide range. In the East, it is a beautiful red or orange snake, with red blotches on a light ground color. In the West, it is usually light gray with brown or dark gray blotches. Two long blotches on the neck meet to form a *spear-point* between the eyes. The Corn Snake climbs well but is more likely to be found on the ground. It may be more common in many areas than it appears, spending much of its time underground, resting or prowling through rodent burrows.

RAT SNAKE **42–101 in.**

This large, handsome serpent comes in many colors. It is black in the Northeast, but elsewhere it is gray, orange, or golden yellow. Some Rat Snakes are plain, others are blotched or have four dark stripes running down their bodies. When cornered, many of these snakes stand up and fight, with the front part of the body raised, the head drawn back in an S, and the mouth held open, ready to strike. They squeeze mice, young rats, or small birds to death in their strong coils. Rat Snakes have angled belly scales that let them climb trees easily. They live in woodlands, swamps, and farmlands. Range includes most of the eastern half of the U.S.

INDIGO SNAKE

CORN SNAKE

three
variations

RAT SNAKE

GLOSSY SNAKE **26–70 in.**

Looks like a pale Bullsnake. Light cream or tan above, with slightly darker blotches. Its washed-out appearance accounts for its nickname of "faded snake." It prefers open areas of the Southwest, such as grassland, sagebrush flats and deserts, usually where there is sandy soil. It is an excellent burrower. In hot weather it stays underground during the day. Eats lizards, snakes, and small mammals, which it may kill by constriction.

BULLSNAKE **36–110 in.**

A large white or yellowish snake with dark blotches and a white to yellowish belly. This snake occurs from coast to coast but goes by several different names: It is the Gopher Snake in the West, the Bullsnake in the central U.S., and in the East it is called the Pine Snake. It is a good climber and burrower, active chiefly by day except during hot weather. When alarmed, it hisses loudly, flattens its head and vibrates its tail, and may strike vigorously. This behavior, along with their markings, causes these non-venomous snakes to be mistaken for rattlesnakes and killed.

COMMON KINGSNAKE **30–82 in.**

Kingsnakes are well known for killing and eating other snakes, even venomous ones. In fact, two of them should be fed in separate cages. Otherwise they may start to eat at opposite ends of the same food animal, and when their heads meet, one snake may engulf the other! They kill their prey by constriction. In the East, Common Kingsnakes are shiny black, with large, bold, white *chain links*. In the West, they usually have alternating *bands* of black and white. They may also be *speckled* or nearly all black. Found in a variety of habitats—woodland, grassland, fields, scrubland, and desert—in most of the southern half of the U.S. and into Mexico.

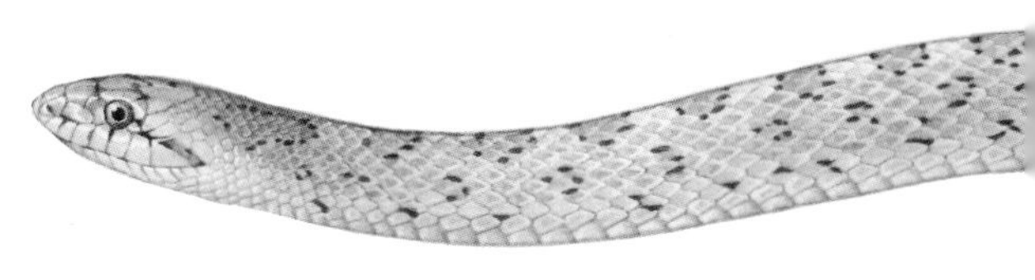

GLOSSY SNAKE

BULLSNAKE

COMMON KINGSNAKE

MILK SNAKE **16–54 in.**
Milk Snakes are tricolored, usually with red, black, and white cross rings or blotches. The reddish parts are bordered by black. The belly usually has black markings. The Milk Snake got its name from the nonsensical belief that it milks cows. In fact, it is one of our most beneficial serpents because it eats mice, often prowling through barns in search of rodents. The Milk Snake's range includes most of the U.S., and the color and shape of its markings vary. In the East, it looks rather like the venomous Copperhead; in the South and West, it may resemble the venomous Coral Snake. This "mimicry" may protect the Milk Snake from predators.

GROUND SNAKE **8–19 in.**
A small, shiny snake that may be hard to identify because its color and pattern are so variable. It may be plain brown or orange, or it may have bands that are black, brown, or orange. Ground Snakes are found most often in the great open spaces of the plains and southwestern states. They are secretive, and are most often discovered by overturning stones, boards, or trash. They eat centipedes, scorpions, spiders, and insects.

WESTERN SHOVELNOSE SNAKE **10–17 in.**
This snake has a *flattish snout*, and its lower jaw is inset within the upper one. This keeps the snake from getting sand in its mouth when "sand swimming," which is more like wriggling through loose sand than tunneling in it. The ground color is whitish or yellow, with black bands, and many individuals also have red bands. A nocturnal reptile that lives in the driest parts of the southwestern desert.

two variations

MILK SNAKE

two variations

GROUND SNAKE

WESTERN SHOVELNOSE SNAKE

SCARLET SNAKE **14–32 in.**

With its red, whitish (or yellow), and black coloring, the Scarlet Snake is a very convincing "mimic" of the dangerous Coral Snakes. Unlike them, however, its *black and red markings touch.* It differs from the Milk Snake by having a plain *white belly.* The Scarlet Snake is a burrower that lives in the southeastern quarter of the country.

LONGNOSE SNAKE **20–41 in.**

Another Coral Snake look-alike with red, black, and sometimes yellow markings. This one is quite speckled with white. The Longnose Snake lives in western deserts and prairies, and is nocturnal. It defends itself very effectively by twisting its body in a way that smears blood, feces, and musk over itself and its captor's hands.

Coral Snakes

Their small mouths and short fangs make it hard for them to bite most parts of the human body (except fingers and toes), but the Coral Snakes' venom affects the central nervous system. Several harmless snakes have the colorful markings of Coral Snakes. Think of a traffic light: yellow means caution, red means stop. If these two warning colors touch on a snake's body, it is poisonous (the Western Shovelnose Snake is an exception).

EASTERN CORAL SNAKE **20–48 in.**

A shiny snake with red, yellow, and black rings, the *red and yellow rings touching.* The secretive Eastern Coral Snake lives in pine woods, pond borders, and jungly hardwood thickets in the southeastern states, including Texas.

WESTERN CORAL SNAKE **13–21 in.**

Also has red, yellow, and black bands, *red and yellow touching.* This snake spends much of its time underground in dry and semiarid habitats. If disturbed, it hides its head under its coils and may raise its tail, tightly coiled in a way that resembles a head.

SCARLET SNAKE

LONGNOSE SNAKE

EASTERN CORAL SNAKE

VENOMOUS

WESTERN CORAL SNAKE

VENOMOUS

Pit Vipers

All our dangerously poisonous serpents except the Coral Snakes belong to this group. The name comes from the deep pit on each side of the head, between the eye and the nostril. The pit is a sensory organ that detects heat, helping the viper take aim when striking at warm-blooded prey. Any snake with such a pit is poisonous, but don't approach live ones in the field close enough to see the pit!

COPPERHEAD **20–53 in.**

This variable snake ranges from Massachusetts to Texas. It usually has a coppery red head and an *hourglass pattern* of dark chestnut bands. The color and the shape of the bands vary with locality. The Copperhead is usually quiet, almost lazy, and would rather flee than fight. Once aroused, however, it strikes vigorously and vibrates its tail, which produces a lively tattoo against whatever it touches. Mice are the staple of the diet. Copperheads are gregarious, gathering in autumn at hibernating dens that are often shared with other species of snakes.

COTTONMOUTH **30–74 in.**

This large snake spends a great deal of time in the water. It is usually olive, brown, or black. A pattern of muted bands is most obvious in younger snakes, sometimes disappearing completely in old adults. Often called "water moccasin," but so are other large, harmless water snakes (see page 96). You can usually tell a Cottonmouth from a water snake by its behavior. Cottonmouths stand their ground or crawl slowly away; water snakes flee quickly or drop with a splash into the water. Cottonmouths vibrate their tails when excited; water snakes do not. A thoroughly aroused Cottonmouth throws its head back and holds its mouth wide open, revealing the white interior that gives it its name. Range extends from Virginia to Texas and Florida.

COPPERHEAD

COTTONMOUTH

Pit Vipers: Rattlesnakes

Rattlesnakes are often heard before they are seen, and their rattle is usually enough to identify them. In the very young, the rattle is just a button. A new segment is added at each shedding time, the chain growing longer as the snake grows older.

MASSASAUGA **16–40 in.**

A *spotted* rattler with a row of large dark blotches down the back and rows of smaller blotches on its sides. Ground color is usually gray, lighter in the western parts of its range. It prefers wet prairies, bogs, and swamps, and is sometimes called the "swamp rattler." Massasaugas are usually mild-mannered, seldom rattling unless they are thoroughly aroused. Their range extends from the Great Lakes to Arizona.

PIGMY RATTLESNAKE **15–31 in.**

A southeastern rattler with a *skinny tail* and a *tiny rattle* that sounds like a buzzing insect. It is usually gray or brown, with dark bars or blotches. The behavior of Pigmys varies; some strike furiously, while others are lethargic and do not even rattle.

MOJAVE RATTLESNAKE **24–51 in.**

Identify this rattlesnake of the southwestern deserts by its *narrow black tail rings*. It is gray, greenish, or brownish with darker blotches on its back. The Mojave Rattlesnake often resembles the Western Diamondback, but its venom is usually much more powerful, making this excitable serpent one of the most dangerous snakes in the United States.

WESTERN RATTLESNAKE **15–65 in.**

The Western Rattlesnake is blotched in shades of brown and black, and its ground color usually matches the surrounding soil. This Rattlesnake has a large western range, and favors habitats of rocky outcrops and ledges. In cooler areas it may gather in large numbers in animal burrows or caves.

MASSASAUGA

PIGMY RATTLESNAKE

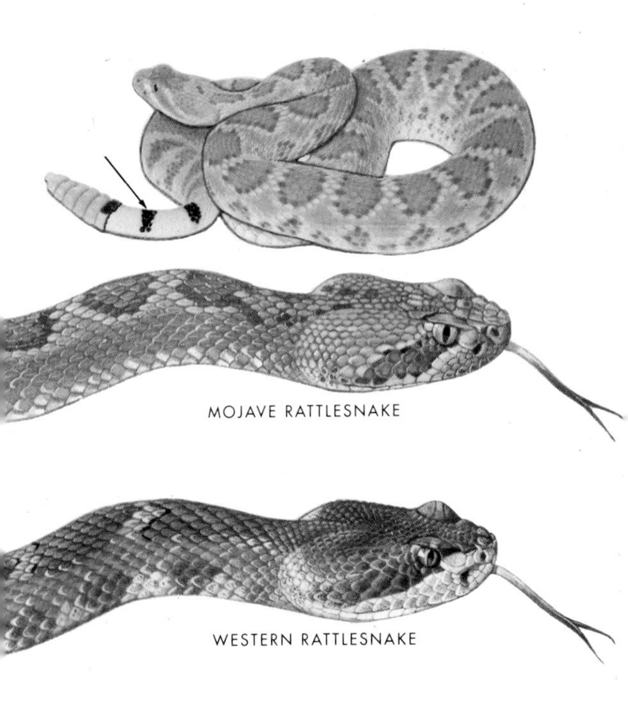

MOJAVE RATTLESNAKE

WESTERN RATTLESNAKE

SIDEWINDER **17–33 in.**

This western desert rattler has *pointed, hornlike scales* over its eyes that may act as sunshades. Although it is chiefly active at night, it ambushes lizard prey by day. It is generally pale-colored with darker blotches down its back. The Sidewinder is named for its S-curved method of locomotion. Sidewinding is a way to move very quickly through open, sandy areas where there is little plant life and no rocks to get in the way. It leaves tracks of parallel J-shaped marks, with the hook pointing in the direction of travel.

TIMBER RATTLESNAKE **36–74 in.**

The only Rattlesnake in most of the Northeast. It is quite variable in color, but usually has black or dark brown crossbands on a ground color of yellow or pale tan. Black ones are not unusual. In the Northeast, Timber Rattlers may gather in dens to hibernate, often with Copperheads and other species. Habitats include forests in the Northeast, thickets in the South, and wooded stream valleys in the Midwest.

EASTERN DIAMONDBACK RATTLESNAKE **33–96 in.**

The compact coils, broad head, and loud, buzzing rattle make this an ominously impressive snake to meet in the field. The dark brown or black *diamonds* are outlined in yellow. Ground color is olive, brown, or black. At home in the dry woodlands of the South, and occasionally swims to outlying Keys from the Florida coast. If approached, many will stand their ground, but if hard pressed will back away, rattling vigorously but still facing the intruder.

SIDEWINDER

TIMBER RATTLESNAKE

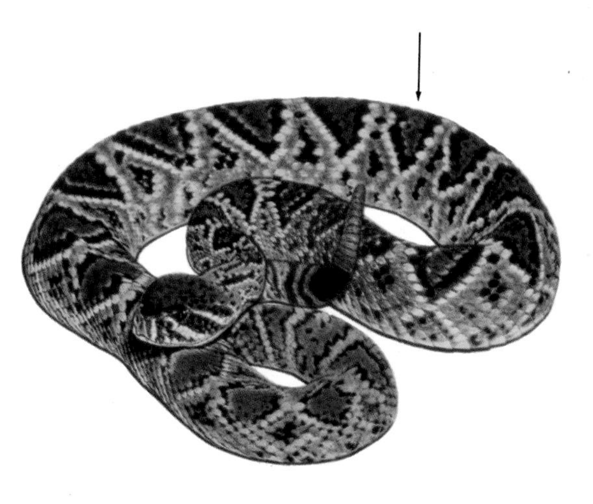

EASTERN DIAMONDBACK RATTLESNAKE

WESTERN DIAMONDBACK RATTLESNAKE **30–84 in.**

The great size, a willingness to stand its ground when threatened, and the loud, buzzing rattle are usually enough to identify this reptile. When fully aroused, like most rattlers it may raise the head and a loop of its neck high above the coils, gaining elevation for aiming and striking. The diamonds often are not clear-cut, and the entire head and body may have a dusty look. Generally brown or gray, with yellowish tones. Tail is strongly ringed with black on white or gray. The Western Diamondback lives in the Southwest, on desert flats as well as in rocky cliffs and mountain canyons. This snake is responsible for more serious snakebites and deaths than any other North American serpent. From the standpoint of sheer size, it and the Eastern Diamondback rank among the world's largest and most dangerous snakes.

WESTERN DIAMONDBACK RATTLESNAKE

VENOMOUS